NASA SP-3003

I0506238

HANDBOOK OF SPACE-RADIATION EFFECTS ON SOLAR-CELL POWER SYSTEMS

By William C. Cooley and Robert J. Janda

Prepared under contract for NASA by Exotech, Incorporated, Alexandria, Virginia, and reproduced photographically from copy supplied by the contractor

Office of Scientific and Technical Information
NATIONAL AERONAUTICS AND SPACE ADMINISTRATION

1963

Washington, D.C.

ACKNOWLEDGEMENTS

The guidance of Walter C. Scott of National Aeronautics and Space Administration Headquarters is gratefully acknowledged. Special thanks are due Kandiah Shivanandan who served as consultant to the authors. The cooperation of many other individuals and organizations in providing data is also acknowledged with thanks.

FOREWORD

In order that photovoltaic solar cells may be used more effectively to provide electrical power for spacecraft, it is desirable to refine the engineering design of solar cells and power systems, based on a better understanding of the space radiation environment and of radiation damage effects on semiconductor components. The objective of this handbook is to provide a summary of some of the useful analytical methods and test data which can be applied in designing radiation resistant power systems. It is found that the most serious obstacle which prevents accurate prediction of solar cell degradation for earth satellites is our inadequate knowledge of the fluxes and energy spectra of electrons and protons in the magnetosphere. When the space environment can be better defined, it will be possible to predict more accurately the degradation of present day types of silicon solar cells, for which the radiation damage characteristics have been quite well determined by laboratory research. As improved types of solar cells and materials become available, it will be necessary to make laboratory measurements of their radiation resistance to protons and electrons in order to predict their performance in space. Therefore, the performance data presented herein may become obsolete. However, it is hoped that the methods for data correlation and analysis presented here will be an aid to understanding the nature of the problems involved, and will be useful in the conduct of engineering analysis.

TABLE OF CONTENTS

I	Introduction	1
II	Theory of the Solar Cell	2
	A. The Photovoltaic Effect	2
	B. Solar Cell Design	3
	C. Solar Cell Performance	5
III	Radiation Damage to Solar Cells	12
	A. Production of Defects and Recombination Centers by Radiation	12
	B. Simplifying Assumptions	13
	C. Effect of Radiation on Diffusion Length	14
	D. Electron Damage Coefficients	18
	E. Proton Damage Coefficients	18
	F. Calculation of the Diffusion Length From Laboratory Data	27
	G. Correlation of the Performance of Solar Cells as a Function of Diffusion Length	28
IV	The Space Radiation Environment	44
	A. Introduction	44
	B. McIlwain's Coordinates	44
	C. Electron Fluxes and Energy Spectra	46
	D. Proton Fluxes and Energy Spectra	51
V	Correlation of Satellite Test Data on Solar Cell Performance	55
VI	Design Methods for Solar Cell Power Systems	63
	A. General Requirements	63
	B. Selection of Type of Solar Cells	63
	C. Degradation Rates in Particular Orbits as a Function of Shield Thickness	63
	D. Calculation of Solar Cell Degradation Rates	64
	E. Sample Calculation of Radiation Damage in Equatorial Orbits	65
	F. Comparison of Calculated and Observed Damage on S-27 Alouette	74

TABLE OF CONTENTS (CONT'D)

Appendix A. Radiation Effects on Solar Cell Cover Slide Materials and Adhesives	A 1
A. Introduction	A 1
B. Damage Mechanism	A 1
C. Cover Slide Requirements	A 2
D. Electron and Proton Damage Experiments	A 2
E. Conclusion	A 14
Appendix B. Space Radiation Effects on Transistors and Diodes	B 1
A. Introduction	B 1
B. Radiation Damage Mechanisms	B 1
C. Surface Effects of Radiation	B 1
D. Degradation of Diffusion Length in Transistors	B 2
E. Selection of Transistor Type	B 5
F. Diodes	B 5
G. Shielding Against Space Radiation	B 6
References	R 1

LIST OF FIGURES

Figure Number		Page
1	N on P Silicon Solar Cell	4
2	Typical Current-Voltage Characteristic for a Solar Cell	6
3A	N/P Silicon Solar Cell Current Density as a Function of Diffusion Length	7
3B	P/N Silicon Solar Cell Current Density as a Function of Diffusion Length	8
4	Short Circuit Current Density vs. Diffusion Length for Two Light Sources	10
5	Maximum Power Output as a Function of Diffusion Length for N/P Solar Cells in Space Sunlight	11
6A	Diffusion Length Degradation of N/P Solar Cells With 1 MEV Electrons	15
6B	Diffusion Length Degradation of P/N Solar Cells With 1 MEV Electrons	16
7	Degradation of Diffusion Length in 1 ohm-cm P-Type Silicon as a Function of Integrated Flux	17
8	Electron Damage Coefficient as a Function of Electron Energy for N/P Silicon Cells	19
9	Electron Damage Coefficient as a Function of Electron Energy for P/N Silicon Cells	20
10	Effect of Electron Energy on Damage Rate for N/P and P/N Silicon Cells	21
11	Proton Damage Coefficient for N/P Silicon Cells as a Function of Proton Energy	25
12	Proton Damage Coefficient for P/N Silicon Cells as a Function of Proton Energy	26
13	Typical Spectral Response of RCA N/P (1 ohm-cm) Solar Cell to 1.0 MEV Electrons	29
14	Current-Voltage Characteristic for BTL Blue-Shifted N/P Silicon Solar Cell	31
15	Short Circuit Current Ratio vs. Diffusion Length Ratio For Silicon Solar Cells	33
16	Degraded vs. Initial Diffusion Length For 75% Short Circuit Current In Silicon Solar Cells	34
17	Typical Variation of Maximum Power Ratio With Short Circuit Current Ratio	36

LIST OF FIGURES (CONT'D)

Figure Number		Page
18	Power vs. Voltage at Various Stages of Degradation For Blue N/P 1 Ohm-cm BTL Silicon Cells	37
19	Maximum Power Ratio vs. Diffusion Length Ratio For Silicon Solar Cells	38
20	Current vs. Voltage Curves For N/P 10 Ohm-cm Silicon Solar Cells at Various Cell Temperatures Before And After 1 MEV Electron Radiation	40
21	Current vs. Voltage Curves For P/N 1 Ohm-cm Silicon Solar Cells At Various Cell Temperatures Before and After 1 MEV Electron Irradiation	41
22	Electrical Characteristics as Functions of Temperature at Increasing Levels of 1 Mev Electron Flux for N/P Solar Cells	42
23	Electrical Characteristics As Functions Of Temperature At Increasing Levels of 1 Mev Electron Flux For P/N Solar Cells	43
24	The B - L Magnetic Coordinate System	45
25	The Mapping Of The Polar Coordinates R and λ Onto the B - L Plane	47
26	The B - L Map of Electron Fluxes Approximately One Week After Starfish	48
27	The R - λ Map Of Electron Fluxes Approximately One Week After Starfish	49
28	The Energy Distribution Of Fission Electrons	50
29	Summary Of Observed Omnidirectional Intensities Of Electrons Obtained From Explorer 12 and 14	52
30	Proton Flux Map	53
31	Flight Data On Short Circuit Current Ratio For Silicon Solar Cells vs. Time In Orbit	58
32	Life Of N/P 1 Ohm-CM Silicon Solar Cells vs. Shield Thickness For Various Satellites	59
33	Life Of P/N Silicon Solar Cells vs. Shield Thickness For Various Satellites	60
34	Composite Graph Of Life Of Silicon Solar Cells vs. Shield Thickness For Various Orbits	62
35	Mass-Range For Protons Through Cover Slide Materials	66
36	Conversion Ratio From Fission Electrons To 1 MEV Electrons	67
37	Damage Rate By Proton Flux As A Function Of Proton Energy	70

LIST OF FIGURES (CONT'D)

Figure Number		Page
38	The Function F(x) Vs. x For Calculating Damage By Protons Below 60 MEV	72
39	Time For 25% Reduction In Short Circuit Current For N/P 1 Ohm-CM Silicon Cells In Circular Equatorial Orbits	75
A-1	Effect of 1 MEV Electron Irradiation On Transmittance Of Solar Cell Cover	A 9
A-2	Effect Of 1 MEV Electron Irradiation On Transmittance Of Solar Cell Cover	A 10
A-3	Effect Of 1 MEV Electron Irradiation On Transmittance Of Solar Cell Cover	A 11
A-4	Effect Of Exposure To 1.2 MEV Electrons On Solar-Energy Transmission Of Silica Windows	A 12
B-1	Effect Of Gamma Radiation On Collector Reverse Current Of Evacuated And Gas-Filled Silicon Transistors	B 3

LIST OF TABLES

III-1	Proton Damage Coefficients For 1Ω-cm P-Type Silicon (N/P Cells)	23
III-2	Proton Damage Coefficients For 1Ω-cm N-Type Silicon (P/N Cells)	24
IV-1	Data On Nuclear Explosions Which Injected Particles Into The Magnetosphere	54
V-1	Orbital Parameters Of Satellites	56
V-2	Estimated Time In Orbit To Degrade Cells To 75% Of Initial Short Circuit Current	57
A-1	Effects Of 1 MEV Electrons On Spectral Transmittance (T) Of Solar Cell Cover Materials	A3
A-2	Effects Of 4.6 MEV Protons On Spectral Transmittance (T) Of Solar Cell Cover Materials	A6
A-3	Effects Of 1.2 MEV Electron Radiation On Transparent Materials	A7

I INTRODUCTION

Bombardment of solar cells by energetic electrons and protons in space produces radiation damage which decreases their power output. Earth satellites which operate within the range of altitudes between about 400 and 40,000 miles may encounter appreciable fluxes of high energy charged particles which are trapped in the earth's magnetic field. In the design of solar cell power systems, consideration should be given to the electron and proton fluxes and their energy spectra along the proposed orbit, the type of solar cells, the material and thickness of shielding material, the solar cell operating voltage, and the expected operating temperature range. This handbook is intended to provide a review of research data and analytical methods which can be used to design radiation resistant silicon solar cell power systems for earth satellites which have orbits passing through the magnetosphere. A discussion of the effects of solar flares on solar cells has not been included. For a discussion of this subject, one may refer to a report by Madey (Ref. 53) who calculated the power degradation expected from a solar flare.

Appendices are included which summarize data on radiation darkening of solar cell cover slide materials and radiation effects to transistors and diodes. Radiation effects to electronic components such as resistors and capacitors are not discussed because the radiation damage expected in the space environment for such non-semiconductor components is generally negligible.

II THEORY OF THE SOLAR CELL

A. The Photovoltaic Effect

An n-type semiconductor crystal contains a small concentration of doping atoms which ionize and supply electrons for conduction. A p-type crystal contains doping atoms which take electrons from the lattice, leaving "holes" which can conduct current by moving from one atom to the next as an adjacent electron moves in.

When a p-n junction is produced in a crystal, electrons from the n-side thermally diffuse into the p-region where they are called minority carriers. Similarly, holes from the p-side become minority carriers in the n-region. These minority carriers rapidly become neutralized by recombining with the majority carriers. Since each region was initially electrically neutral, the loss of electrons from the n-region and the loss of holes from the p-region each contributes to establishing a potential difference across the junction with a plus voltage on the n-side. This built in electric field can be used for photovoltaic power generation.

When a photon of light is absorbed in a crystal it will ionize an atom and release an electron, thereby producing an electron-hole pair. After a short burst of illumination, the electrons and holes which have been injected, both diffuse until they find an opposite number and recombine. The length of time for a minority carrier to recombine (the lifetime) depends on the density of recombination centers, which are crystal defects which provide sites where minority carriers are captured and then recombined readily with majority carriers.

The distance travelled by a minority carrier before recombining is called the minority carrier diffusion length, L, which is related to the lifetime, τ, by:

$$L = \sqrt{D\tau} \qquad (1)$$

where D is the diffusion constant. In silicon, the value of D is about 38 cm^2/sec. for electrons and 13 cm^2/sec. for holes at 300° K. Both L and τ are dependent on the impurity content and on the crystal perfection, which is affected by radiation damage. They also are functions of the temperature.

When light is absorbed in the region near the p-n junction of a solar cell, which is usually 0.25 to 1 micron from the surface, some of the electrons and holes will diffuse to the junction. At this point, the charges will be separated by the built in electric field at the junction, thereby providing a current which can flow through an external load. This method for direct conversion of photon energy to electrical power is known as the photovoltaic effect.

The current output of a solar cell under red light, which produces electron-hole pairs several hundred microns below the junction, depends on the magnitude of the diffusion length for minority carriers in the base region. Therefore exposure to radiation which produces recombination centers, primarily affects the response of a solar cell to red light.

B. Solar Cell Design

A typical solar cell consists of a rectangular wafer of high purity semiconductor crystal with a p-n junction formed near the front surface and electrical connections applied to the front layer and to the base region, as shown in Fig. 1. Usually a conducting grid is used on the front surface to reduce the internal resistance to lateral current flow in the thin surface layer.

Until recently most solar cells in the United States were made with a p-type layer on n-type silicon. The development of radiation resistant n-on-p silicon cells by Mandelkorn (Ref. 54) has lead to their selection for many satellites.

The improved performance of n-on-p cells under radiation can be attributed partly to the fact that electrons, which are the minority carriers in the p-type base material have about a three times greater diffusion constant than holes do in the n-type base of p-on-n cells. A contributing factor is that the types of recombination centers produced by radiation in p-type silicon are less effective in shortening the minority carrier diffusion length than those produced in n-type silicon.

Optimization of solar cell design for radiation resistant performance requires locating the p-n junction only about 0.25 micron below the front surface to maximize the collection of current generated by photons near the blue end of the visible spectrum, which are absorbed and produce electron-hole pairs very near the surface. Cells designed in this way are called shallow-diffused or blue-shifted cells. They derive a

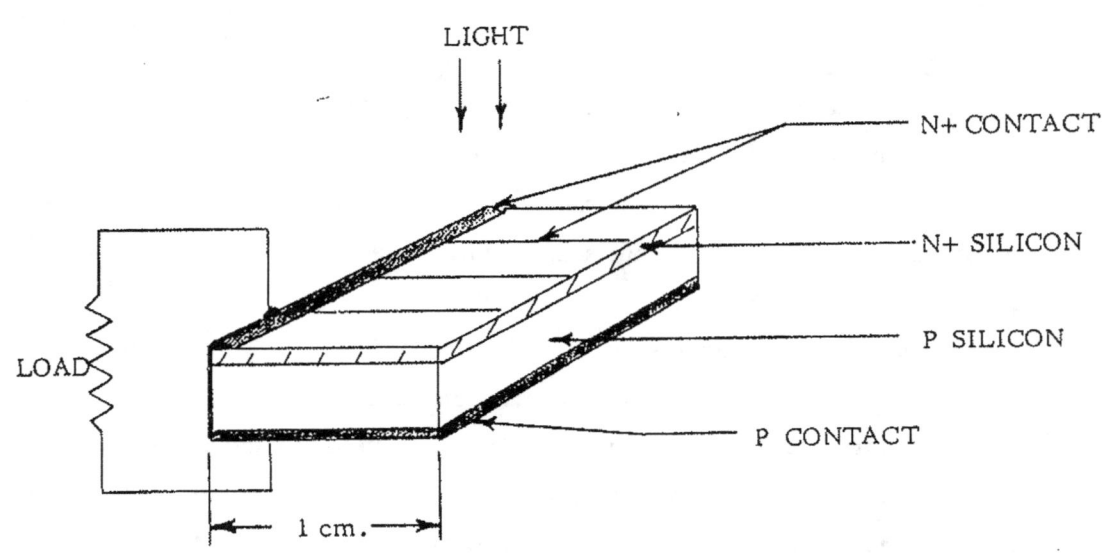

FIG. 1 N ON P SILICON SOLAR CELL

smaller fraction of their power from the red end of the solar spectrum and are therefore less sensitive to degradation of the diffusion length in the base.

C. Solar Cell Performance

The diode equation for a solar cell is:

$$I = I_L - I_0 \left[e^{\frac{qV}{AKT}} - 1 \right] \qquad (2)$$

where:

I = current through load

I_0 = saturation current of the p-n junction

I_L = light generated current

V = voltage across the load

q = electronic charge (esu)

K = Boltzmann's constant

A = a constant normally between 1 and 3

T = temperature ($^\circ$K)

This equation predicts a current voltage characteristic as shown in Fig. 2

Under short circuit conditions, the influence of I_0 is eliminated and the short circuit current is equal to I_L. The light-generated current is a function of the light intensity, the absorption coefficient for photons, the geometry of the cell and the diffusion length for minority carriers in the n and p regions.

Kleinman (Ref. 48) has analyzed the effect of diffusion length on the short circuit current. His theoretical solution indicates a variation of the short circuit approximately proportional to the logarithm of the diffusion length over the range from L = 10 to 200 microns which is of practical importance.

It has been found experimentally by Space Technology Laboratories (Ref. 28) that data on short circuit current as measured under tungsten light can be correlated with the logarithm of diffusion length for typical p/n and n/p cells. These data are shown in Fig. 3A and 3B

Experiments by Bell Telephone Laboratories have yielded data which permitted computation of the short circuit current under space sunlight as a function of diffusion

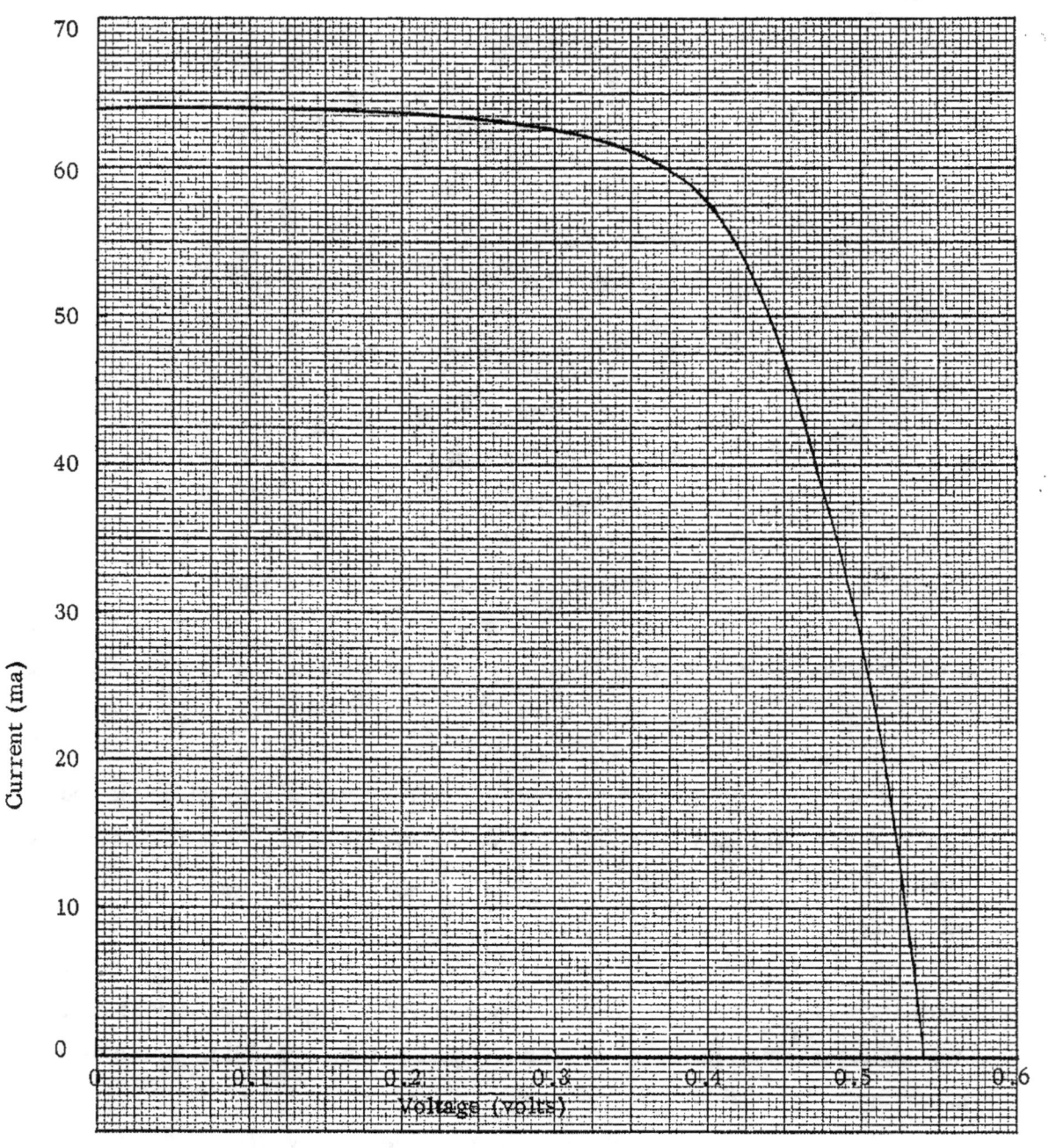

FIG. 2 TYPICAL CURRENT - VOLTAGE CHARACTERISTIC FOR A SOLAR CELL

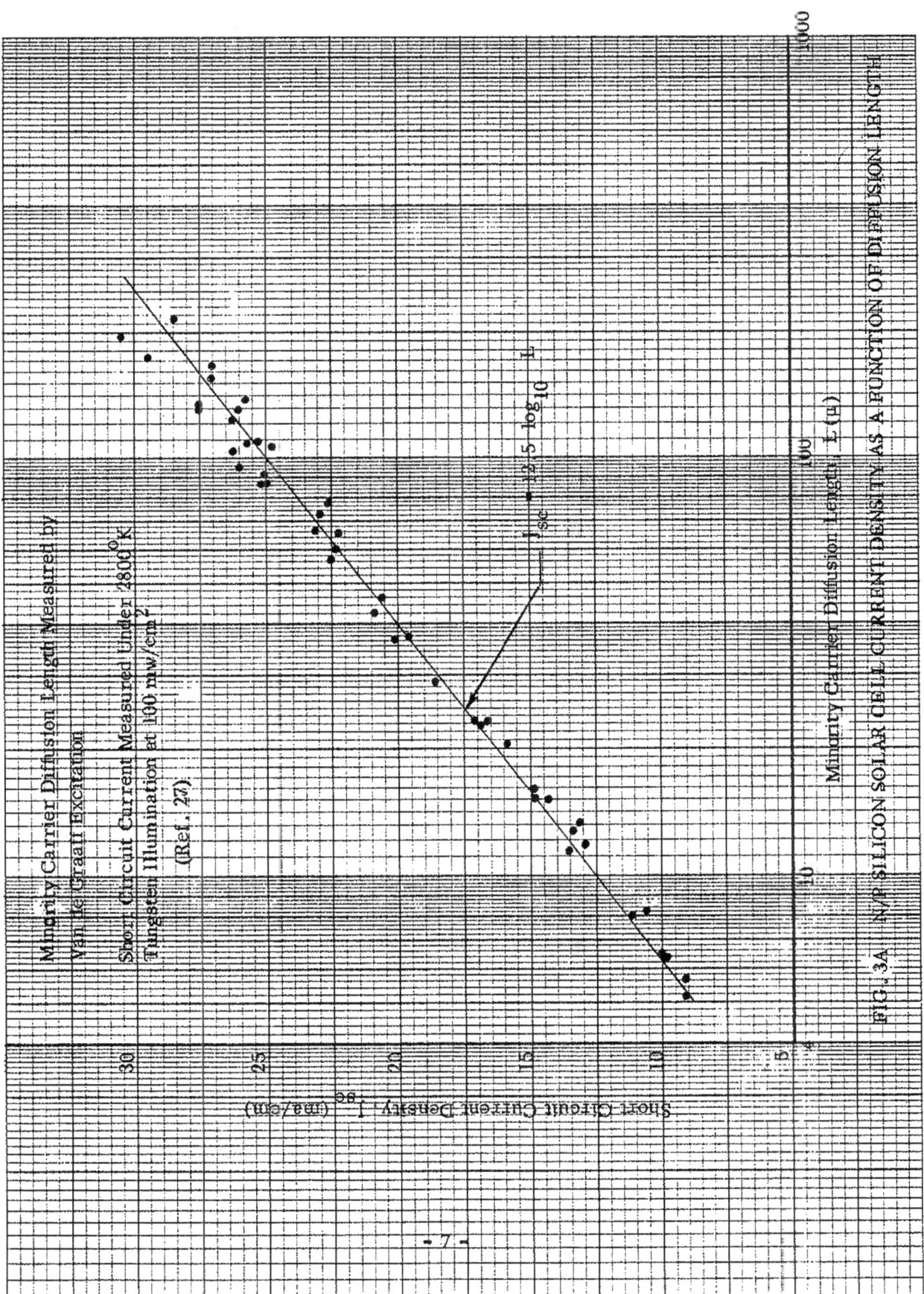

FIG. 3A N/P SILICON SOLAR CELL CURRENT DENSITY AS A FUNCTION OF DIFFUSION LENGTH

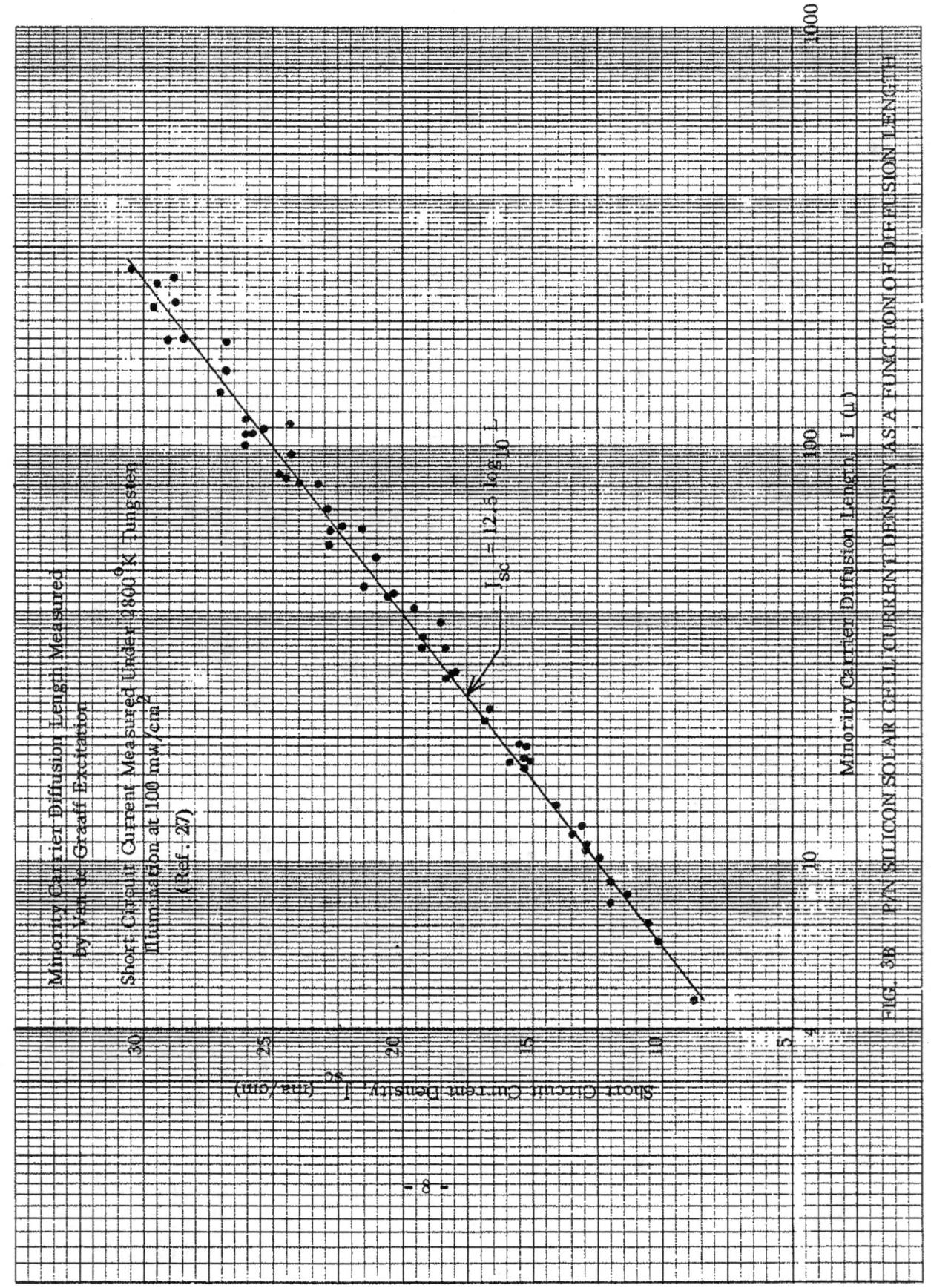

FIG. 3B P/N SILICON SOLAR CELL CURRENT DENSITY AS A FUNCTION OF DIFFUSION LENGTH

length, which are shown in Fig. 4 for comparison with the STL correlation line. It is noted that there is some departure from linearity with log L for the BTL data on blue-shifted p/n and n/p cells.

A very significant factor to be noted in Fig. 4 is that the short circuit current does not decrease as rapidly with diffusion length reduction under space sunlight as when the measurement is made under tungsten light, which is excessively rich in red light.

Madey (Ref. 53) has used BTL data (Ref. 75) to correlate maximum power under space sunlight with the logarithm of diffusion length, as shown in Fig. 5.

Based on the empirical evidence, it is assumed that the prediction of solar cell performance in a radiation environment can be reduced to a calculation of the degradation of diffusion length, coupled with reference to data which show how the performance parameters of the particular type of cell in space sunlight will vary with diffusion length.

The use of minority carrier diffusion length (or lifetime) to correlate radiation damage effects is recommended because it is an accurately measurable quantity (see next section) which is sensitive to damage by penetrating radiation and is independent of other solar cell parameters like surface optical reflectivity, surface recombination velocity, and junction depth, all of which influence measurements of current and power.

In nearly all space applications a glass, silica or sapphire cover slide is used over the cell to aid in radiative heat rejection and to carry a spectrally selective filter as well as to provide radiation shielding and protection in ground handling. Since the efficiency of a solar cell varies (logarithmically) with light intensity, it is desirable to use a cover slide material and adhesive (if necessary) which do not darken under space radiation, or to estimate the decrease in transmittance expected in the wavelength range of interest (from 0.35 to 1.1 microns). Pertinent test data are given in Appendix A.

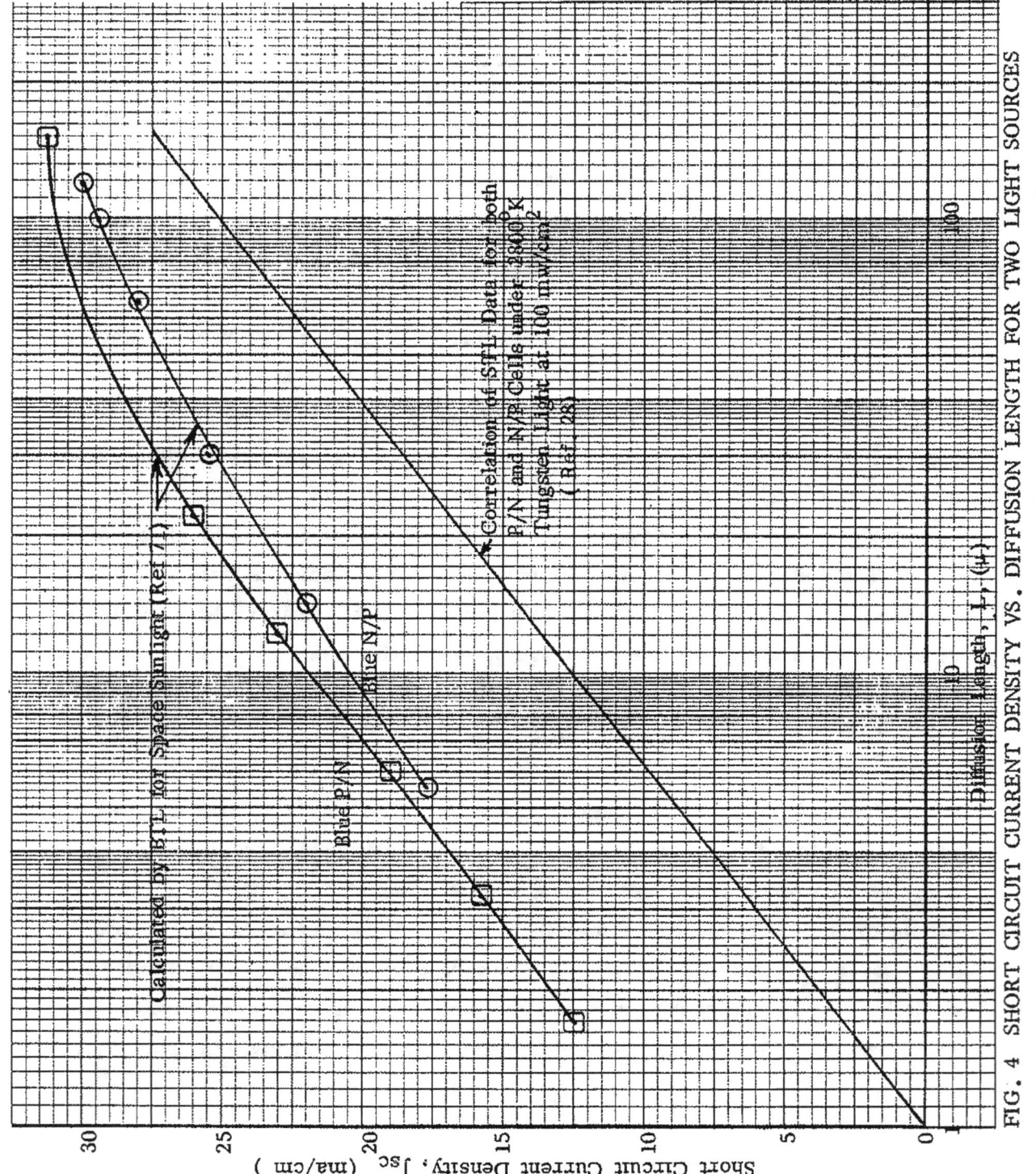

FIG. 4 SHORT CIRCUIT CURRENT DENSITY VS. DIFFUSION LENGTH FOR TWO LIGHT SOURCES

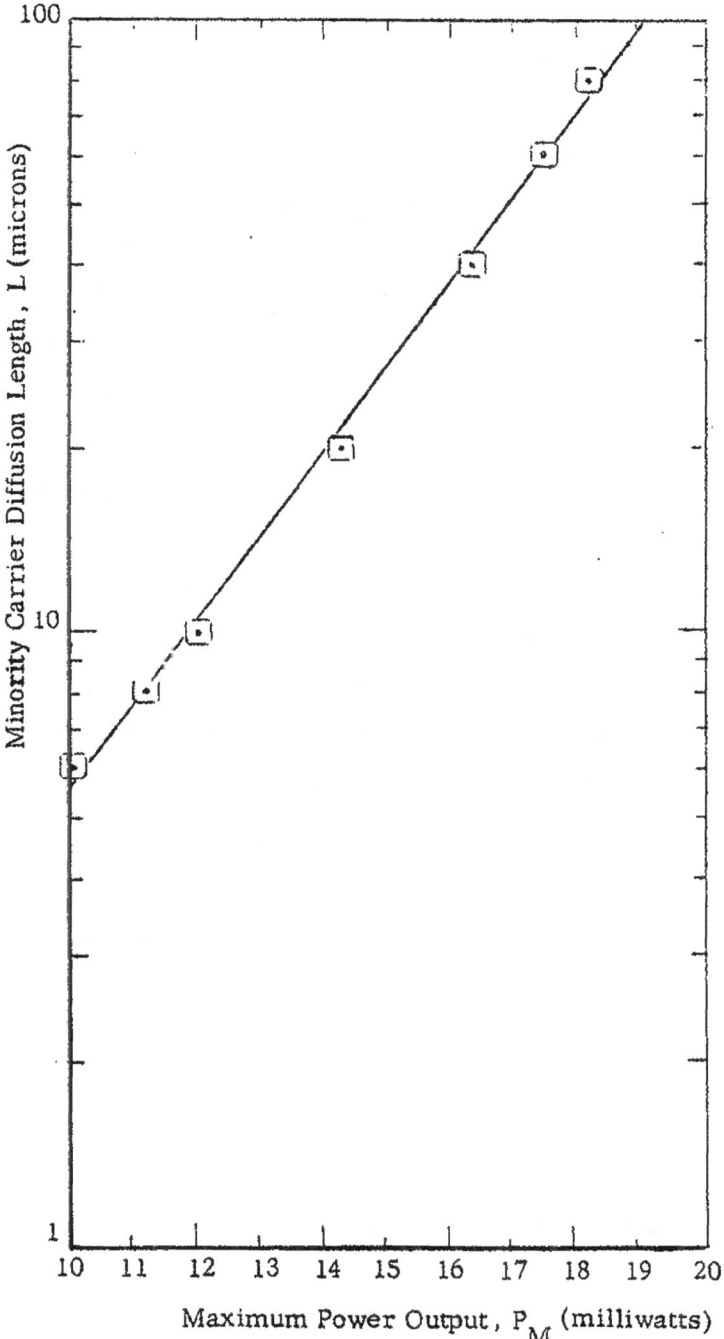

FIG. 5 MAXIMUM POWER OUTPUT AS A FUNCTION OF DIFFUSION LENGTH FOR N/P SOLAR CELL IN SPACE SUNLIGHT

III RADIATION DAMAGE TO SOLAR CELLS

This section includes a discussion of the physics of radiation damage, the effect of radiation on minority carrier diffusion length, analysis of the degradation of diffusion length in space using laboratory data on damage coefficients, and shows how the solar cell parameters such as short circuit current, maximum power and current at a fixed voltage can be correlated with the diffusion length.

A. Production of Defects and Recombination Centers by Radiation

When electrons at energies greater than 145 Kev and protons at energies greater than 98 ev bombard a silicon crystal, they can displace an atom from the crystal lattice, producing a lattice vacancy and a recoil atom which comes to rest as an interstitial atom. With high energy electrons or protons, the recoil atom may have enough energy to displace other atoms before coming to rest. The displacement of a single atom is called a Frenkel defect. The displacement of two adjacent atoms is called a divacancy. When very high energy protons (500 Mev) interact with a nucleus they can cause spallation (star production) which results in several nuclear fragments which stop at short range in the crystal, displacing many atoms.

The crystal defects produced may not be thermally stable except at very low temperatures. At room temperature the vacancies as well as crystal impurity atoms can migrate and form stable defects, some of which act as recombination centers for electrons and holes. The recombination process occurs as the defect center first captures a minority carrier and subsequently captures a majority carrier, thereby annihilating an electron-hole pair. The silicon A-center is a stable defect consisting of a substitutional oxygen atom occupying a site in the silicon lattice. The A-center concentration is dominant in determining the diffusion length of minority carriers (electrons) in p-type silicon when it is irradiated by low energy electrons. P-type silicon is the base material in radiation resistant n/p solar cells and less susceptible to the formation of recombination centers than n-type silicon.

The silicon E-center appears to be a vacancy next to a substitutional phosphorus atom and is produced predominantly in floating zone n-type silicon. Research is continuing to understand the nature of other types of defect centers in semiconductors and to determine whether it is possible to intentionally dope the material in such a way that the stable defects do not act as recombination centers which decrease the diffusion length and lifetime (Ref. 79).

The effectiveness of recombination centers in shortening the diffusion length in silicon is generally not directly proportional to the number of atoms which are displaced by radiation, but depends on the types of defects that are introduced by protons and electrons of different energies, which affect the types and concentrations of the recombination centers that are formed by interaction with impurities (Wysocki, Ref. 88, has shown that a simpler situation prevails in gallium arsenide, where the damage produced by radiation appears to be proportional to the number of atoms displaced.)

The rather wide scatter in the susceptibility of various types of p/n cells to radiation damage is believed to be caused by inadequate control of impurities in the n-type base material which affect the formation of recombination centers.

One effective way to improve the collection efficiency in spite of the presence of recombination centers is to use a "drift-field" solar cell in which the doping is graded to spread out the electrostatic field region at the junction. By this technique (Ref. 20), minority carriers from a larger volume of the cell are electrostatically swept to the junction and are less susceptible to capture by the recombination centers that are present.

B. Simplifying Assumptions

An assumption which is of great value in simplifying the analysis of solar cells is that all the performance parameters of a particular design of solar cell are unique functions of the minority carrier diffusion length in the base region. This is equivalent to the assumption that the only effect of penetrating radiation on a solar cell is to reduce the minority carrier diffusion length in the base region.

This basic assumption is found to be valid when the change in diffusion length is produced by penetrating particles which generate a fairly uniform concentration of defects throughout the front 100 to 200 micron thick layer of a solar cell. In general, protons above a few Mev in energy and electrons above a few hundred Kev in energy impinging on bare solar cells approximately meet this requirement. However, low energy protons which only penetrate a few microns or less will primarily damage the front layer and junction region and the assumption above will be invalid. This more complicated situation is not treated further here because it is not believed to be of great importance when cover shields are used. However, it is pertinent in understanding the behavior of bare solar cells when subjected to low energy protons in space.

C. Effect of Radiation on Diffusion Length

The measurement of diffusion length in a solar cell has in many cases been made by injecting carriers at a constant low rate with penetrating ionizing radiation (e.g. with 1 Mev electrons according to the method by Rosenzweig, Ref. 70) and measuring the short circuit current produced, which is proportional to the diffusion length. Rosenzweig used a 12 mil aluminum foil in the electron beam to produce fairly uniform ionization in the solar cell. A reproducibility of measurements within $\pm$ 3% was achieved using 1 Mev electrons.

Measurements of diffusion length have been made before, during and after bombardment of solar cells with energetic electrons and protons in particle accelerators such as Van de Graaf machines and cyclotrons in order to measure radiation damage.

Experimentally it is found that the minority carrier diffusion length (L) measured at a standard temperature varies with integrated flux of penetrating charged particles ($\bar{\Phi}$) according to:

$$\frac{1}{L^2} = \frac{1}{L_0^2} + K\bar{\Phi} \qquad (3)$$

The damage coefficient K is a function of the type and energy of the particles and of the material and its impurity concentrations. K is by definition equal to $d(1/L^2)/d\bar{\Phi}$ and is the change in $(1/L^2)$ introduced per unit integrated flux (particle/cm^2). It may be called the damage per particle (if L is in cm instead of microns, K has the units of particles^{-1}).

Fig. 6A, 6B, & 7 show empirical correlations of the diffusion length for various types of p/n and n/p solar cells as a function of the integrated flux of 1 Mev electrons (Ref. 28). It is seen that after initial degradation to a point where $\frac{1}{L_0^2} << \frac{1}{L^2}$

the data can be correlated by the approximate equation: $\frac{1}{L^2} \cong K\bar{\Phi}$ or $L \cong (K\bar{\Phi})^{-1/2}$ (4)

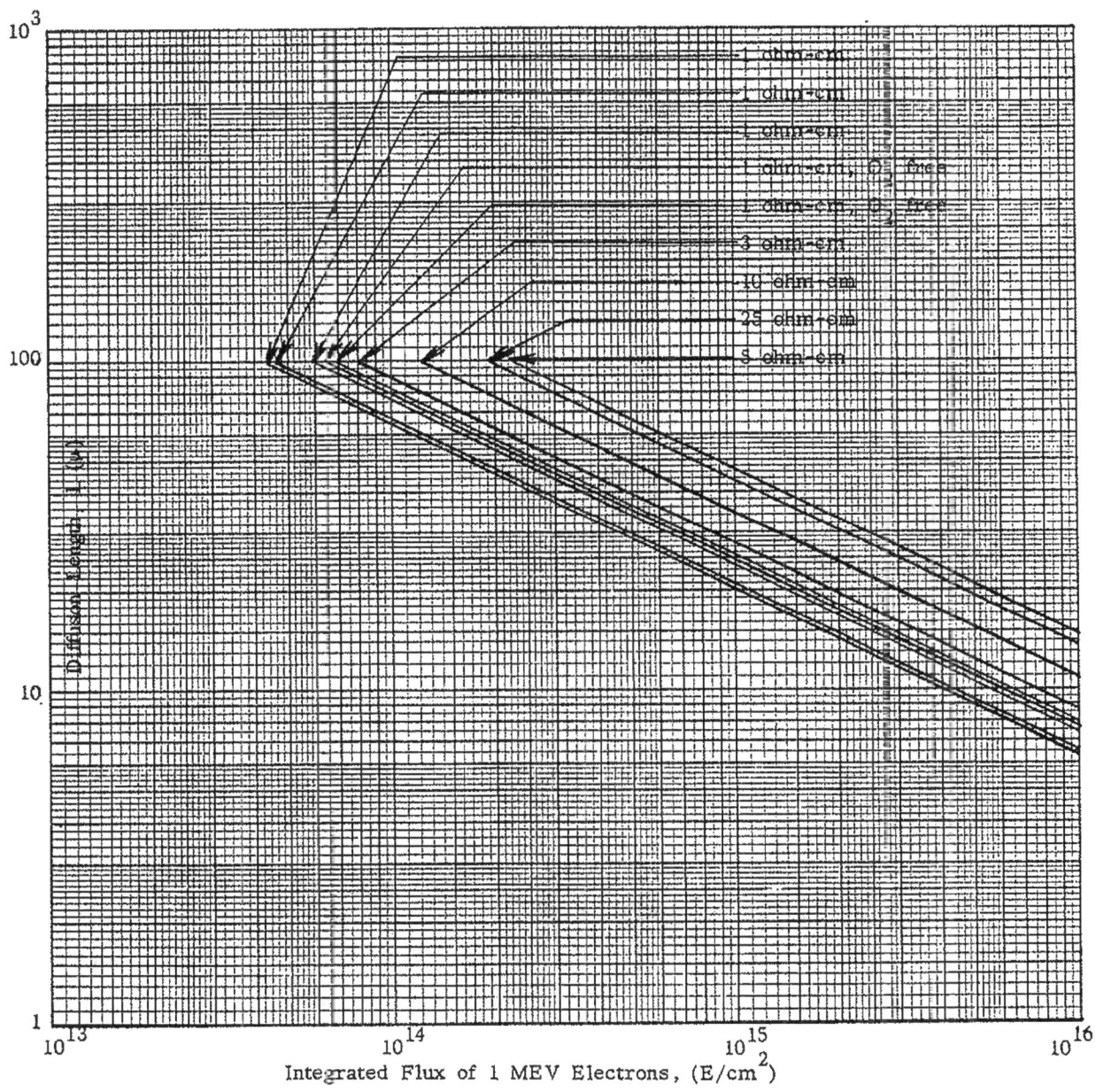

FIG. 6A DIFFUSION LENGTH DEGRADATION OF N/P SOLAR CELLS WITH 1 MEV ELECTRONS

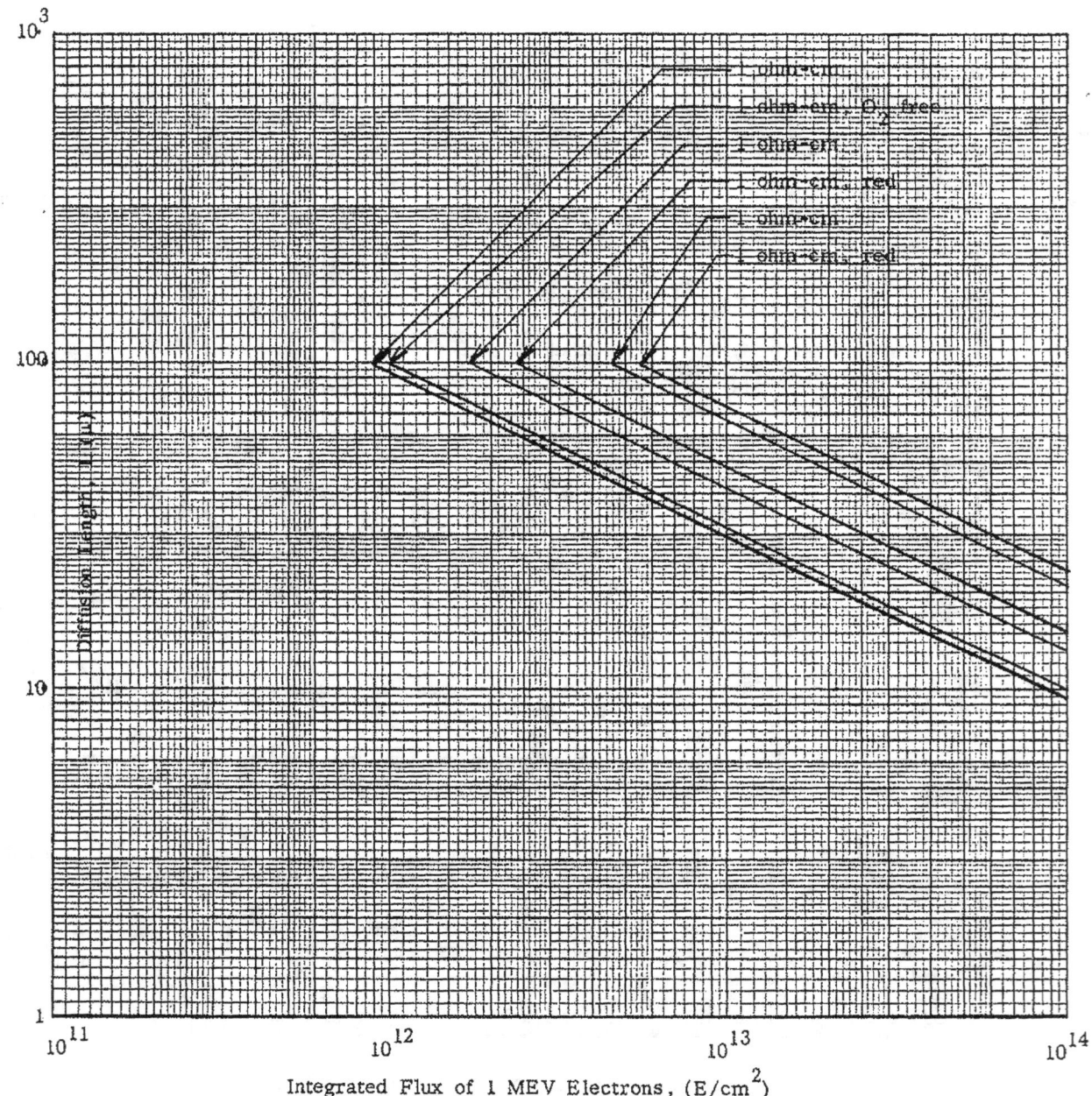

FIG. 6B DIFFUSION LENGTH DEGRADATION OF P/N SOLAR CELLS WITH 1 MEV ELECTRONS

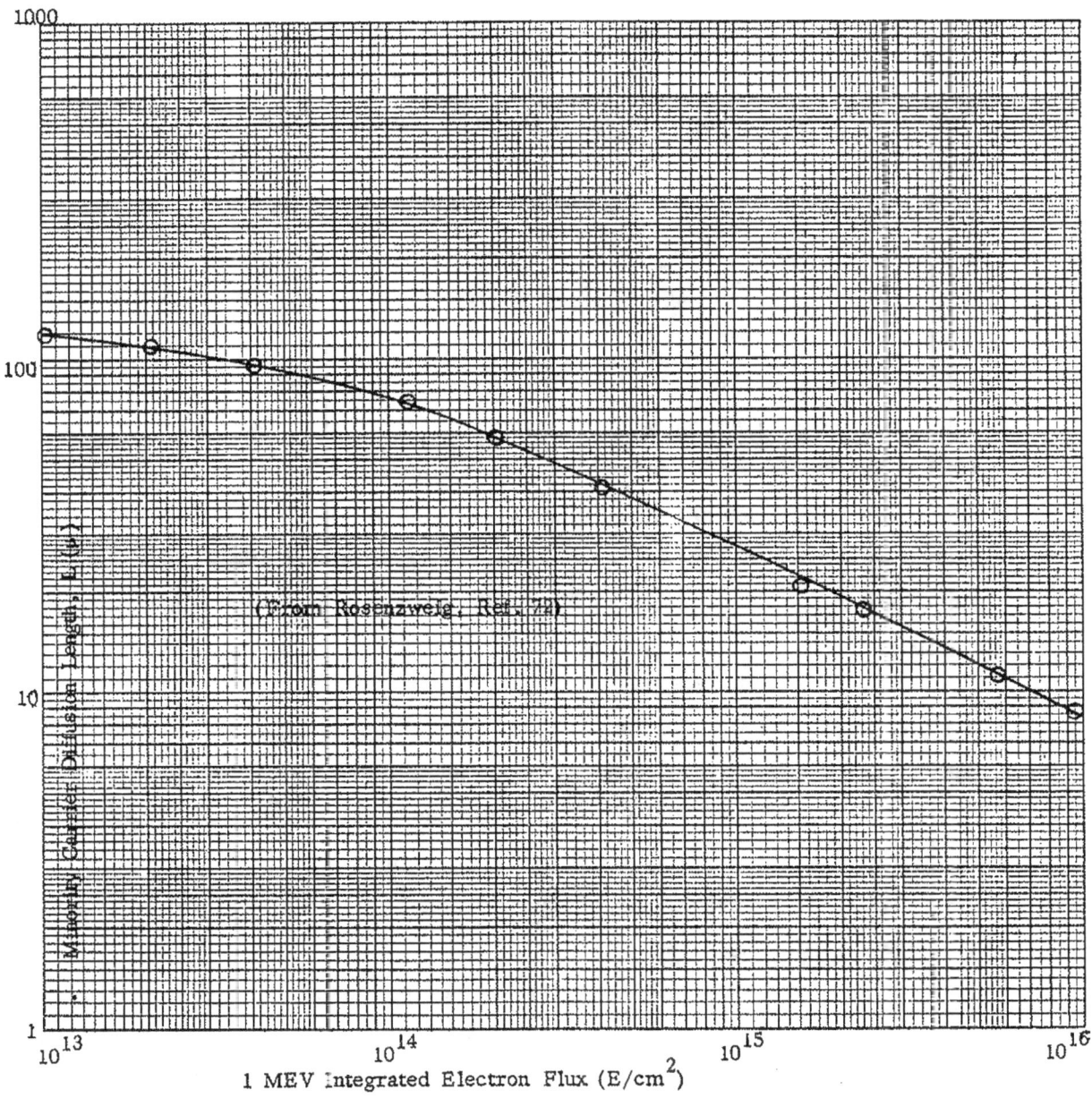

FIG. 7 DEGRADATION OF DIFFUSION LENGTH IN 1 OHM-CM P-TYPE SILICON AS A FUNCTION OF INTEGRATED FLUX

In Fig. 6A it is seen that the data for n/p cells from three manufacturers using 1 ohm-cm p-type silicon for the base material all cluster quite closely together. The improved radiation resistance of cells with higher base resistivity (higher purity) is also clear, with one possible exception.

In Fig. 6B it is noted that p/n cells exhibit a large scatter in the data, presumably because of variations of the impurity content in the base material. (However, Weller (Ref. 83) has shown that oxygen content does not control the recombination of minority carriers in p/n cells when irradiated with 4.8 Mev protons).

D. **Electron Damage Coefficients**

The electron damage coefficients for p/n and n/p cells under Mev electrons have been reported by Rosenzweig, Gummel, and Smits (Ref. 71), and the effect of electron energy has been presented by Rosenzweig (Ref. 72). Damage coefficients for 1 Mev electrons have also been calculated from test data by Denney (Ref. 28) and Statler (Ref. 77). Fig. 8 and 9 summarize the data available. It is seen in Fig. 8 that there is a close verification of the test data at 1 Mev for n/p cells of several base resistivity values among the various sources. However, Fig. 9 shows that there is a wide range of K_E values for p/n cells. As shown by Rosenzweig (Ref. 72) the damage coefficient for p/n cells can be correlated with the particular type of silicon used as the base material. He has also compared the relative damage rates at various electron energies for three types of cells, as shown in Fig. 10. These data indicate that electrons above 1 Mev in energy do an anomalously large amount of damage to the p-type silicon in n/p cells. This effect has also be found by Denney at STL. Wysocki (Ref. 88) has suggested that the high energy electrons (and protons) produce divacancy defects which lead to a more deleterious recombination center in p-type silicon than in n-type.

E. **Proton Damage Coefficients**

Downing and Denney (Ref. 29) have shown evidence that the measured diffusion length and lifetime apparently increase with the minority carrier density (injection level), for material irradiated by high energy protons. Van Lint et al (Ref. 80) have observed that this phenomenon does not seem to be as predominant for 1 Mev electron-irradiated material as it is for material irradiated by high energy protons or higher energy electrons (which produce more complex defects). The question is not yet settled

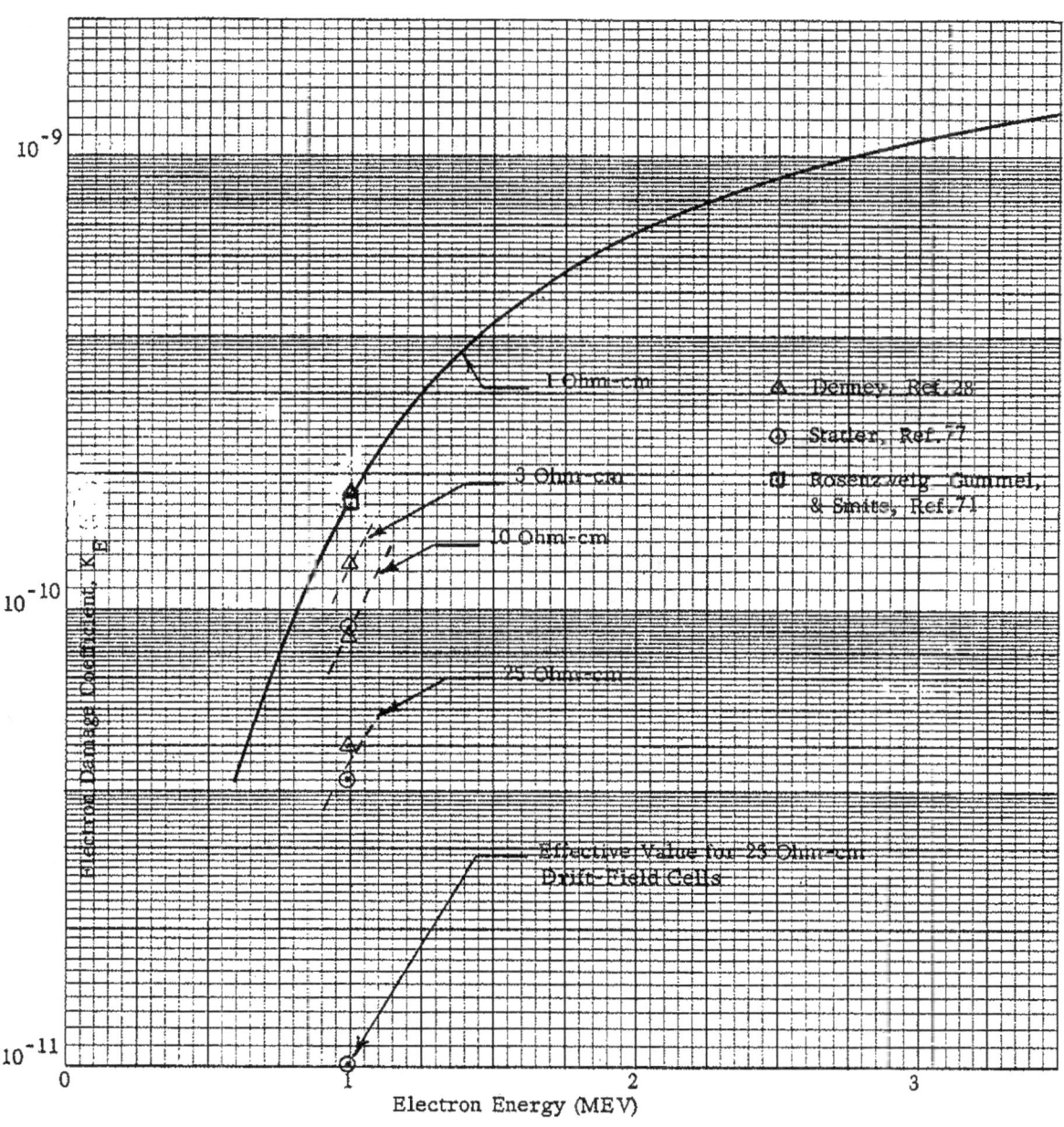

FIG. 8   ELECTRON DAMAGE COEFFICIENT AS A FUNCTION OF ELECTRON ENERGY FOR N/P SILICON CELLS

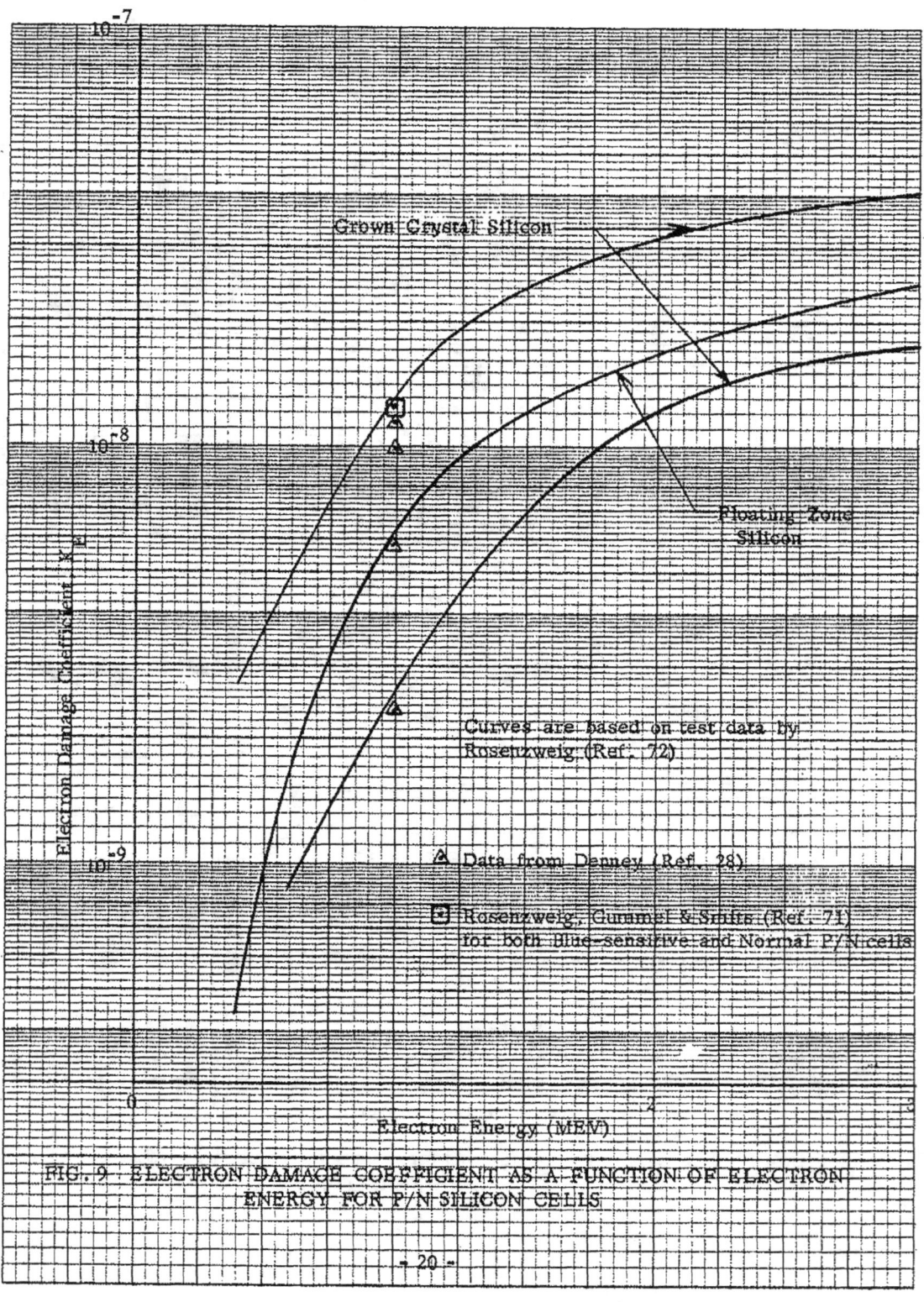

FIG. 9 ELECTRON DAMAGE COEFFICIENT AS A FUNCTION OF ELECTRON ENERGY FOR P/N SILICON CELLS

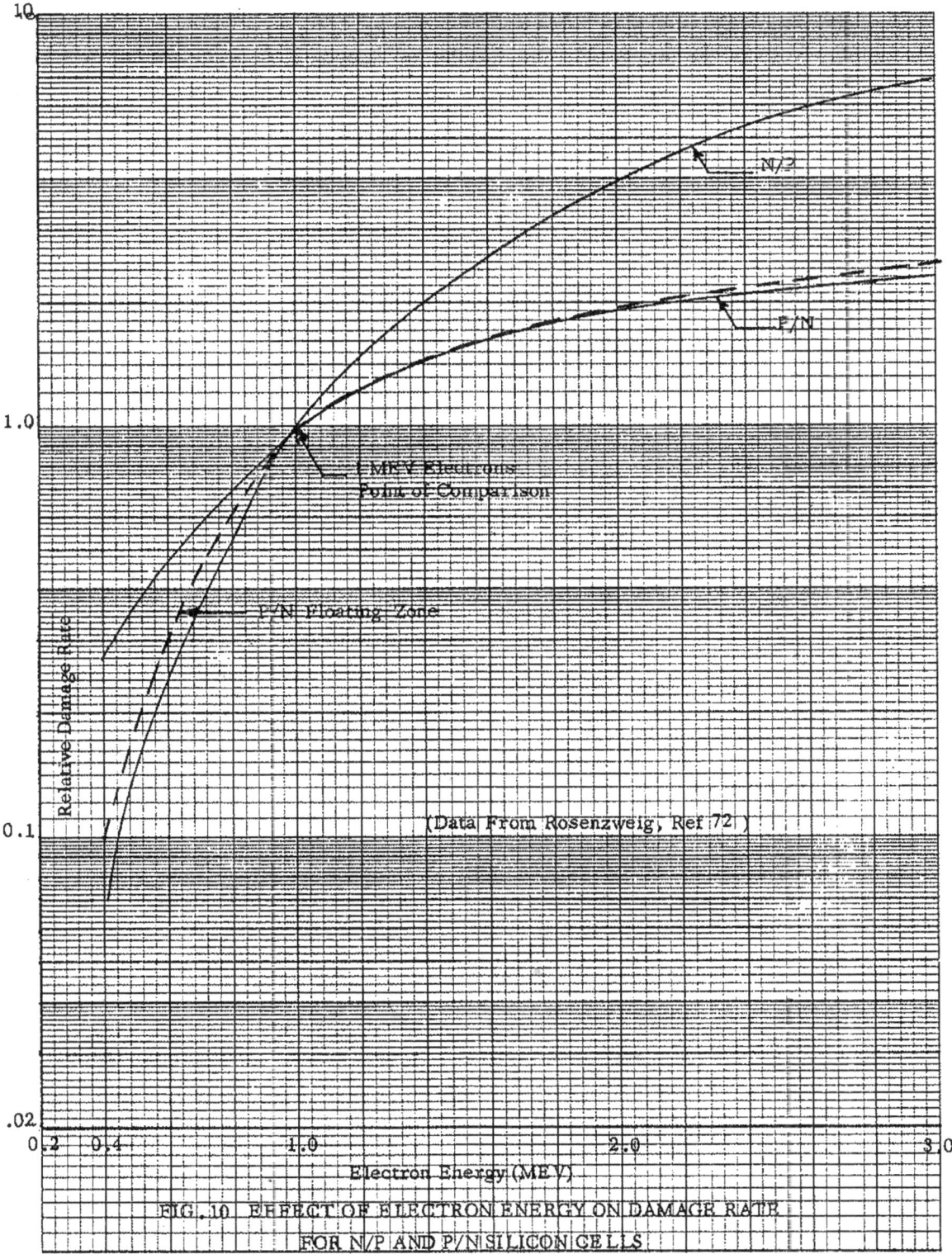

FIG. 10 EFFECT OF ELECTRON ENERGY ON DAMAGE RATE FOR N/P AND P/N SILICON CELLS

as to whether the effect of injection level on measured diffusion length for proton irradiated specimens is dependent on the energy of the protons.

A technique for making measurements using a chopped electron beam in combination with a steady light source to increase the injection level has been developed by Downing and Denney (Ref. 29). However, the fact that the light source does not inject minority carriers uniformly through the solar cell makes interpretation of the results rather difficult. Except where noted below, the values of proton damage coefficients have all been based on measurements of diffusion length using low injection levels.

Measurements of the effect of degradation of diffusion length on the short circuit current and maximum power of solar cells under proton bombardment have been made by several experimenters. In order to correlate the data on a common basis, it was necessary in some cases to compute the damage coefficient K_p from data on the degradation of short circuit current or of maximum power obtained under a variety of light sources. (See Section III-G-3 for the correlation method used).

Tables III-1 and III-2 show the resulting values of the proton damage coefficients at various proton energies for n/p and p/n cells, respectively. These data have been plotted in Fig. 11 and 12. Despite considerable scatter in the data, approximate correlation curves have been drawn under the assumption that the damage coefficient varies inversely with energy up to a certain energy and then remains constant. This correlation is chosen for its mathematical simplicity in damage calculations and may only be accurate within a factor of two or three in the range from 20 to 60 Mev which is usually of considerable importance in determining proton damage rates in space. The points marked with vertical arrows which fall considerably above the correlation line may be in error because the diffusion lengths were measured at low injection levels. By using data by Denney (Ref. 27) the points at 80 Mev and 95.5 Mev can be corrected to correspond to an injection level near that produced by space sunlight, in which case they fall close to the correlation line.

By comparing Fig. 11 and 12, it is seen that the correlation lines yield a ratio of damage coefficients for 1 ohm-cm n-type silicon (p/n cells) to that for 1 ohm-cm p-type silicon (n/p cells) of four (4) at proton energies up to 60 Mev.

TABLE III-1

Proton Damage Coefficients for 1Ω-cm p-Type Silicon (n/p Cells)

Proton Energy MEV	Damage Coefficient, K_p	Reference
1.8	3.2×10^{-6}	6
4.8	1.7×10^{-6}	83
8.3	2.1×10^{-6}	7*
16.8	8.3×10^{-7}	70
17.6	1.0×10^{-6}	6
10.0	9.2×10^{-7}	7*
20.5	4.0×10^{-7}	23**
45.0	1.0×10^{-6}	57
80.0	6.0×10^{-7}	57
95.5	2.4×10^{-7}	23
95.5	5.4×10^{-7}	23
120	4.0×10^{-7}	57
130	3.3×10^{-7}	70
450	1.8×10^{-7}	23
750	2.8×10^{-7}	23

* K_p was calculated from the authors' values of L_0, L and Φ.

** The Φ_c's from six types of cells were averaged and converted to a K value.

TABLE III-2

Proton Damage Coefficients for 1Ω-cm n-Type Silicon (p/n Cells)

Proton Energy, MEV	Damage Coefficient, K_p	Reference
1.8	1.0×10^{-5}	6
4.8	8.7×10^{-6}	83
8.3	1.2×10^{-5}	7*
16.8	5.1×10^{-6}	70
17.6	5.9×10^{-6}	6
20.5	7.9×10^{-7}	23
45.0	6.0×10^{-6}	57
80.0	4.0×10^{-6}	57
95.5	3.0×10^{-7}	23
95.5	2.26×10^{-6}	23
130	2.0×10^{-6}	70
450	3.1×10^{-7}	23
450	1.2×10^{-6}	57
740	4.2×10^{-7}	23

*K_p was calculated from the authors' values of L_0, L and Φ.

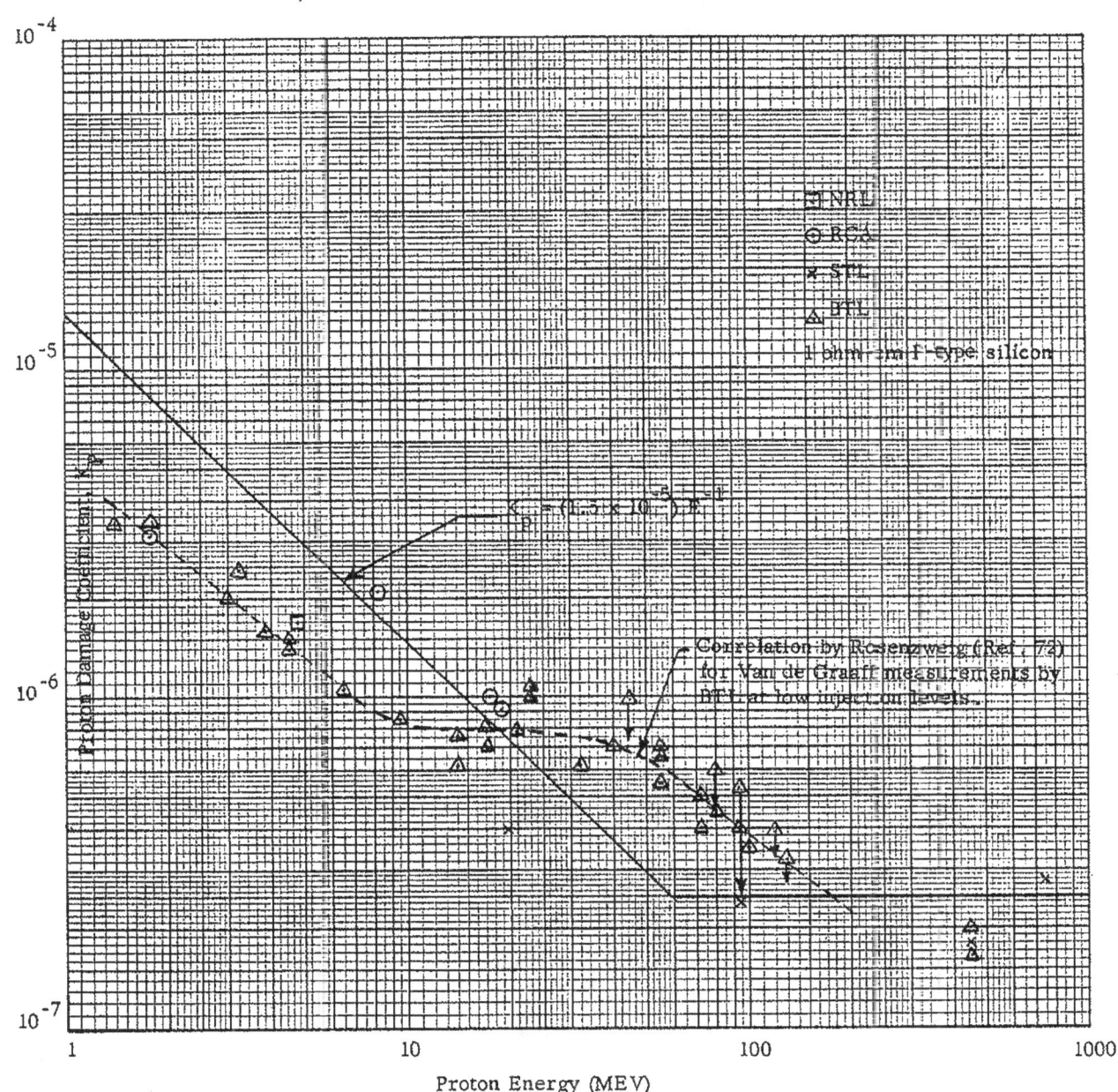

FIG. 11   PROTON DAMAGE COEFFICIENT FOR N/P SILICON CELLS AS A FUNCTION OF PROTON ENERGY

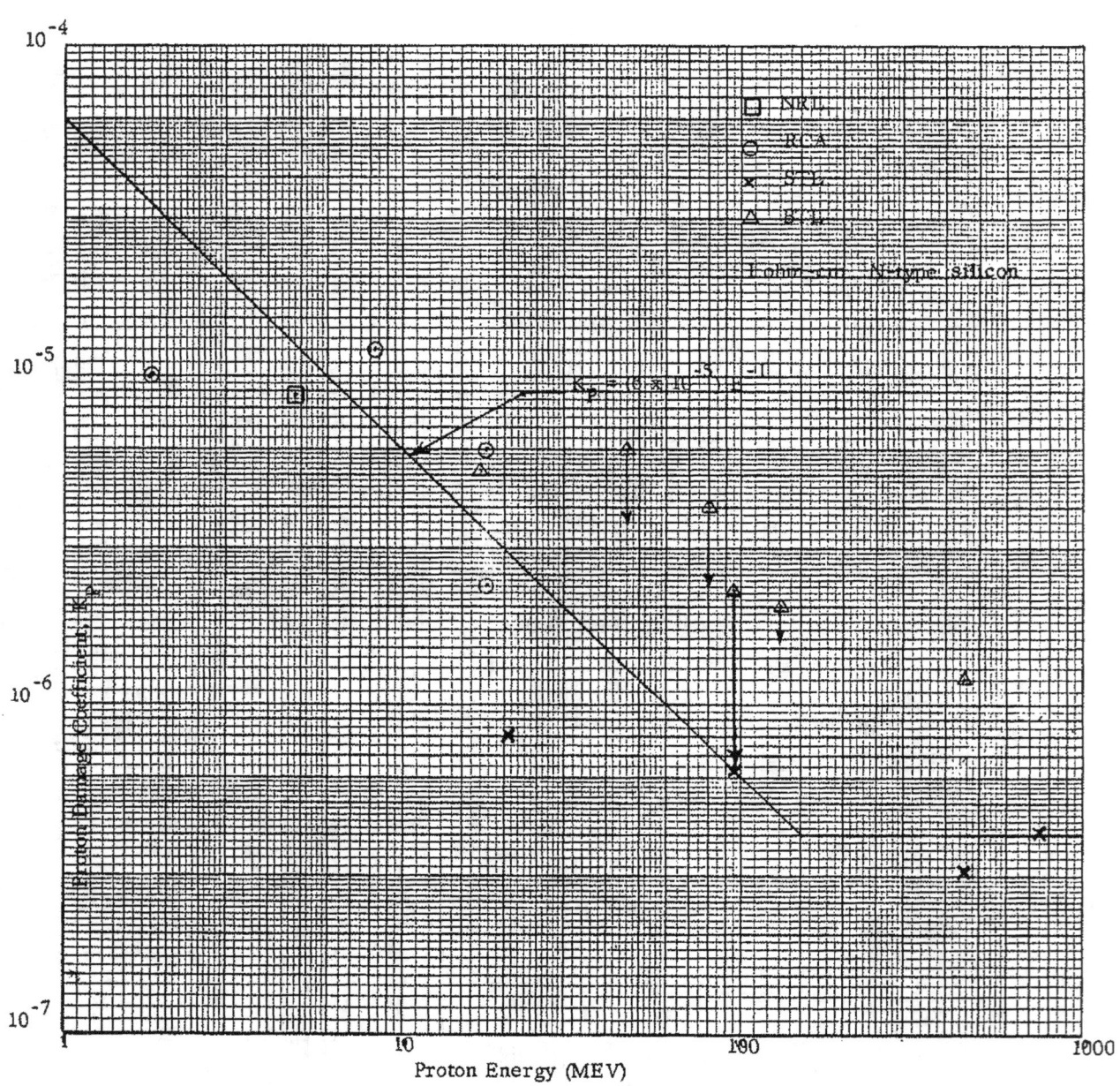

FIG. 12 PROTON DAMAGE COEFFICIENT FOR P/N SILICON CELLS AS A FUNCTION OF PROTON ENERGY

The damage coefficients at proton energies of 1 Mev and lower are only of interest for bare solar cells, which degrade very rapidly when exposed to geomagnetically-trapped protons.

The proton damage coefficients as a function of proton energy E may be approximated by the following relations:

For 1 ohm-cm p-type silicon (n/p cells):
$$K_p = 1.5 \times 10^{-5} E^{-1} \text{ for } 1 < E < 60 \text{ Mev}$$
$$K_p = 2.5 \times 10^{-7} \text{ for } E > 60 \text{ Mev}$$

For 1 ohm-cm n-type silicon (p/n cells):
$$K_p = 6 \times 10^{-5} E^{-1} \text{ for } 1 < E < 150 \text{ Mev}$$
$$K_p = 4 \times 10^{-7} \text{ for } E > 150 \text{ Mev}$$

F. Calculation of the Diffusion Length From Laboratory Data

If more than one type or energy of bombarding particles is present, it is a good approximation to assume that the damage (the differential change in $1/L^2$) is additive and independent of the rate of irradiation. It is also usually assumed that the damage is independent of the temperature and the operating condition of the solar cell during irradiation, although it is known that radiation effects can anneal out at elevated temperature.

Under these assumptions, when several types of particles are present simultaneously or in sequence:

$$\frac{1}{L^2} - \frac{1}{L_0^2} = K_1 \Phi_1 + K_2 \Phi_2 + \tag{5}$$

If the radiation in space has a continuous energy spectrum, it is convenient to define a damage integral by the equation:

$$D = \frac{1}{2} \int_E \int_0^t \rho(E,t) K(E) \, dE \, dt = \left(\frac{1}{L^2} - \frac{1}{L_0^2}\right) \tag{6}$$

where $\rho(E,t)$ = omnidirectional flux per unit energy interval in space (particles/cm^2-sec-Mev) at time t, but modified by the shield over the solar cell.

$K(E)$ = damage coefficient at energy E.

The factor of 1/2 is introduced to account for the usual assumption that there is infinite shielding on the back side of the solar cell.

A laboratory irradiation to a time integrated flux ϕ_1 with particles of energy E_1 is equivalent to a space irradiation if

$$K_1 \phi_1 = \frac{1}{2} \int_E \int_0^t \rho(E,t) K(E) \, dE \, dt \qquad (7)$$

where K_1 is the damage coefficient for particles of energy E_1.

If E_1 is taken as 1 Mev electrons, Equation 7 defines the 1 Mev unidirectional normal incident integrated flux ϕ_1 which is equivalent to a certain irradiation for a time t in space.

For shielded solar cells, the spectral flux $\rho(E,t)$ and the corresponding damage coefficient K should be based on the flux and energy spectrum present behind the shield at the surface of the cell, if the damage coefficient K has been measured for bare cells. Also it is assumed that the particles penetrating the shield have a range longer than about 200 microns and produce uniform damage in the sensitive base region of the solar cell.

The equivalence between doses (integrated fluxes) of particles at any arbitrary energy E_n and the dose at energy E_1 can be established by the relation:

$$K_1 \phi_1 = K_n \phi_n = \left[\frac{1}{L^2} - \frac{1}{L_0^2} \right] \qquad (8)$$

or:
$$\phi_1 = \left(\frac{K_n}{K_1}\right) \phi_n \qquad (9)$$

The ratio (K_n/K_1) is the relative damage rate for particles of energy E_n as compared to energy E_1.

G. <u>Correlation of the Performance of Solar Cells as a Function of Diffusion Length</u>

1. Spectral Response

Fig. 13 shows the change in spectral response of a typical n/p 1 ohm-cm cell under 1 Mev electron bombardment. It is seen that the major degradation occurs in the response near the red end of the spectrum, where photons are absorbed more deeply in the cell and minority carriers must diffuse a greater distance through the damaged material to reach the junction. The value of the minority carrier diffusion length corresponding to each value of integrated flux is shown on the figure.

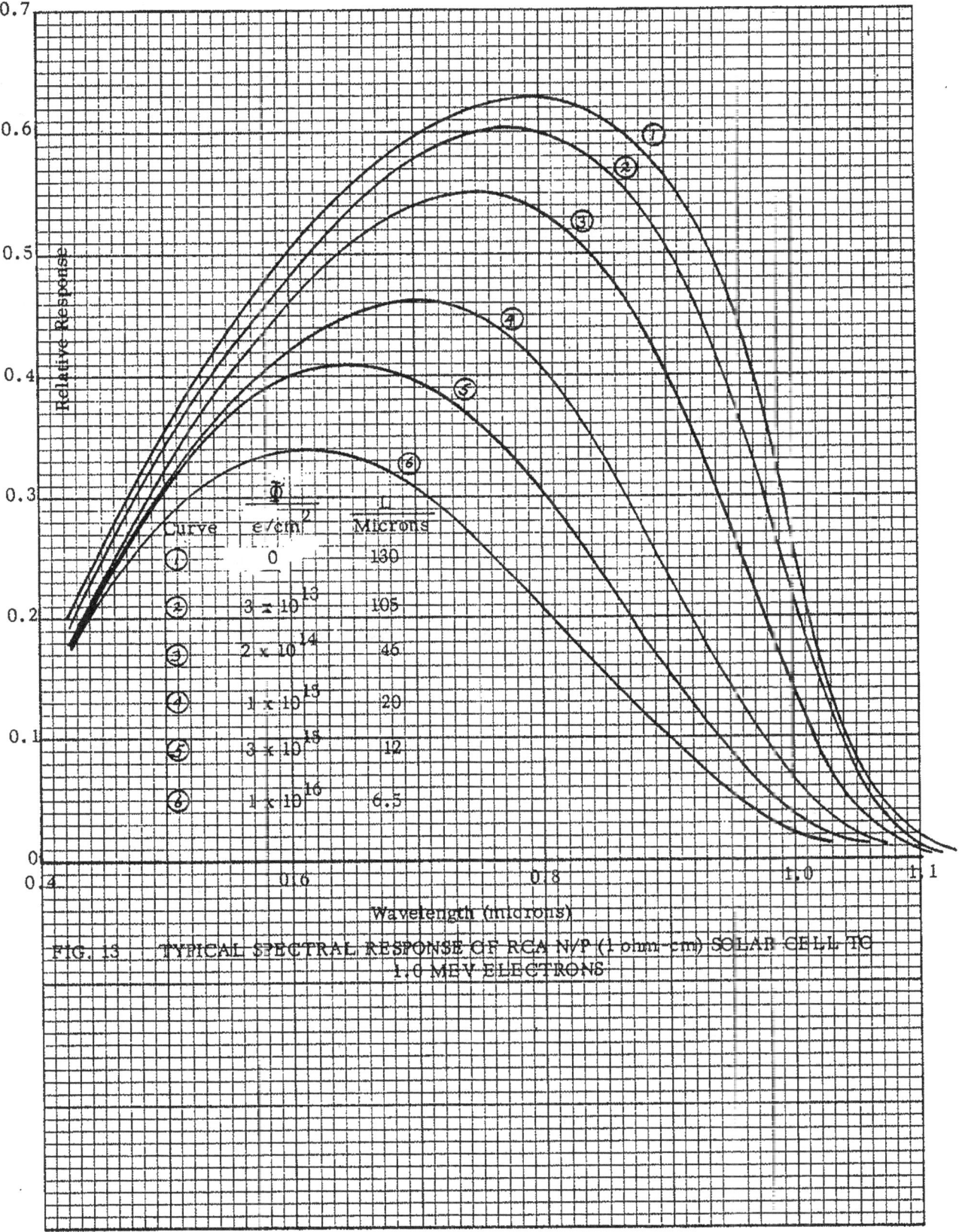

FIG. 13   TYPICAL SPECTRAL RESPONSE OF RCA N/P (1 ohm-cm) SOLAR CELL TO 1.0 MEV ELECTRONS

2. Current - Voltage Characteristic

Fig. 14 is a typical plot showing the change in the current-voltage (i-V) characteristics of a solar cell after various degrees of irradiation. It is seen that the short circuit current degrades more rapidly than the open circuit voltage

3. Short Circuit Current

It was shown in Fig. 3A, 3B, and 4 that the short circuit current and maximum power of solar cells can be correlated with the diffusion length. However, the particular correlation obtained depends on the spectral distribution and intensity of the light source used, and the presence or absence of an anti-reflection coating on the cell (which typically increases the power output by at least 25%, Ref. 75). Also, the initial value of short circuit current or power obtained from a particular cell will be a function of the initial diffusion length of the particular piece of silicon from which it was made, as well as of differences in surface recombination properties and other properties due to manufacturing tolerances.

If one assumes reasonably close control of the manufacturing process, then it should be possible to select an average value for the initial diffusion length and corresponding average values of the initial short circuit current and maximum power as measured under a particular light source which are representative of the particular type of production cells. It is then desirable to predict what the cell performance will initially be in space sunlight and how the performance parameters will degrade as the diffusion length is reduced by radiation.

In order to obtain a generally applicable procedure for predicting performance, the following method was used:

If we assume that short circuit current (I) for a particular type of solar cell varies directly with the logarithm of the diffusion length (L), then one can write:

$$I = C \ln \left(\frac{L}{L'}\right) \tag{6}$$

where C and L' are constants which depend only on the light source.

Then the short circuit current ratio is:

$$\frac{I}{I_0} = \frac{\ln(L/L')}{\ln(L_0/L')} \tag{7}$$

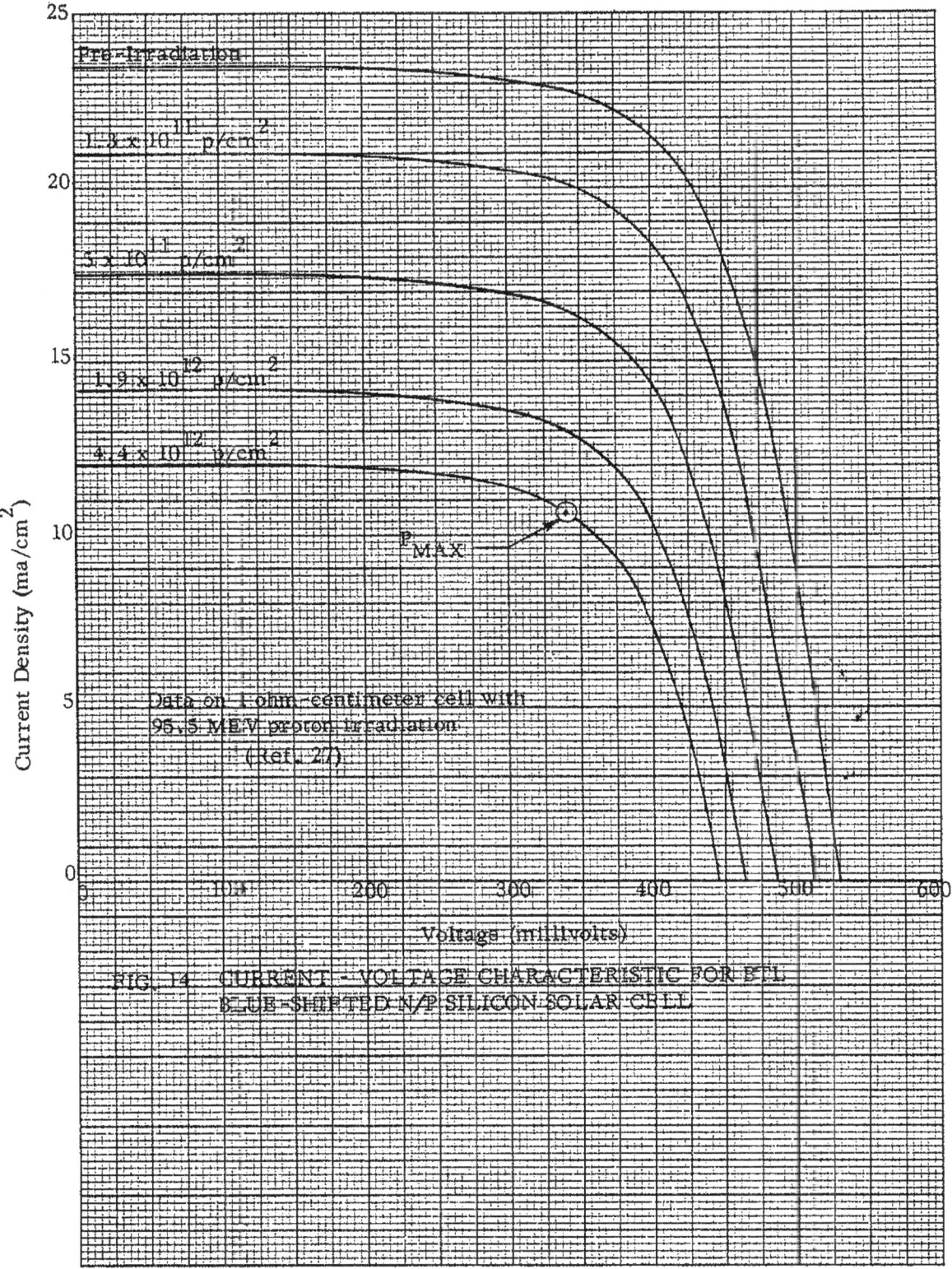

FIG. 14 CURRENT - VOLTAGE CHARACTERISTIC FOR BTL BLUE-SHIFTED N/P SILICON SOLAR CELL

which can also be written as:

$$\frac{I}{I_0} = 1 + \left[\ln\left(\frac{L_0}{L'}\right)\right]^{-1} \left[\ln\left(\frac{L}{L_0}\right)\right] \tag{8}$$

This equation indicates that for a given light source, which determines L', and with a given initial diffusion length L_0, the term $\ln(L_0/L')$ is a constant and the short circuit current should degrade linearly with the term $\ln(L/L_0)$.

Fig. 15 shows the variation of short circuit current ratio with the logarithm of the ratio of final to initial diffusion length for three different light sources. The curves shown are based on measurements for both p/n and n/p cells by Rosenzweig, Gummel and Smits (Ref. 71), Denney, Downing and Van Atta (Ref. 28) and Weller (Ref. 83).

Data from Weller (Ref. 83) for measurements with water-filtered tungsten light show that Equation 8 is approximately satisfied. Data computed for space sunlight (Ref. 71) appear to depart somewhat from linearity as may be seen in Fig. 15.

However, the use of the dimensionless parameter (L/L_0) to correlate test data is still useful as a technique for predicting short circuit current ratio. Using the curves in Fig. 15 it is only necessary to know L_0 to predict the degradation of short circuit current ratio with L. It is believed that the curves in Fig. 15 will permit estimating the short circuit current ratio for blue-shifted silicon solar cells corresponding to any value of L from 10 to 200 microns with reasonable accuracy, independent of whether the cell is n/p or p/n, with base resistivity values from 1 to 10 ohm-cm, and with initial diffusion lengths from 100 to 200 microns.

The problem of predicting the absolute value of initial short circuit current under space sunlight from ground measurements with other light sources is not considered here. (See Ref. 71)

By cross-plotting from Fig. 15, the value of the diffusion length required to yield a 25% reduction in short circuit current under various light sources is plotted as a function of the initial diffusion length in Fig. 16.

From the value of L determined from Fig. 16, the change in $(1/L^2)$ required to produce 25% degradation in short circuit current under various light sources can be

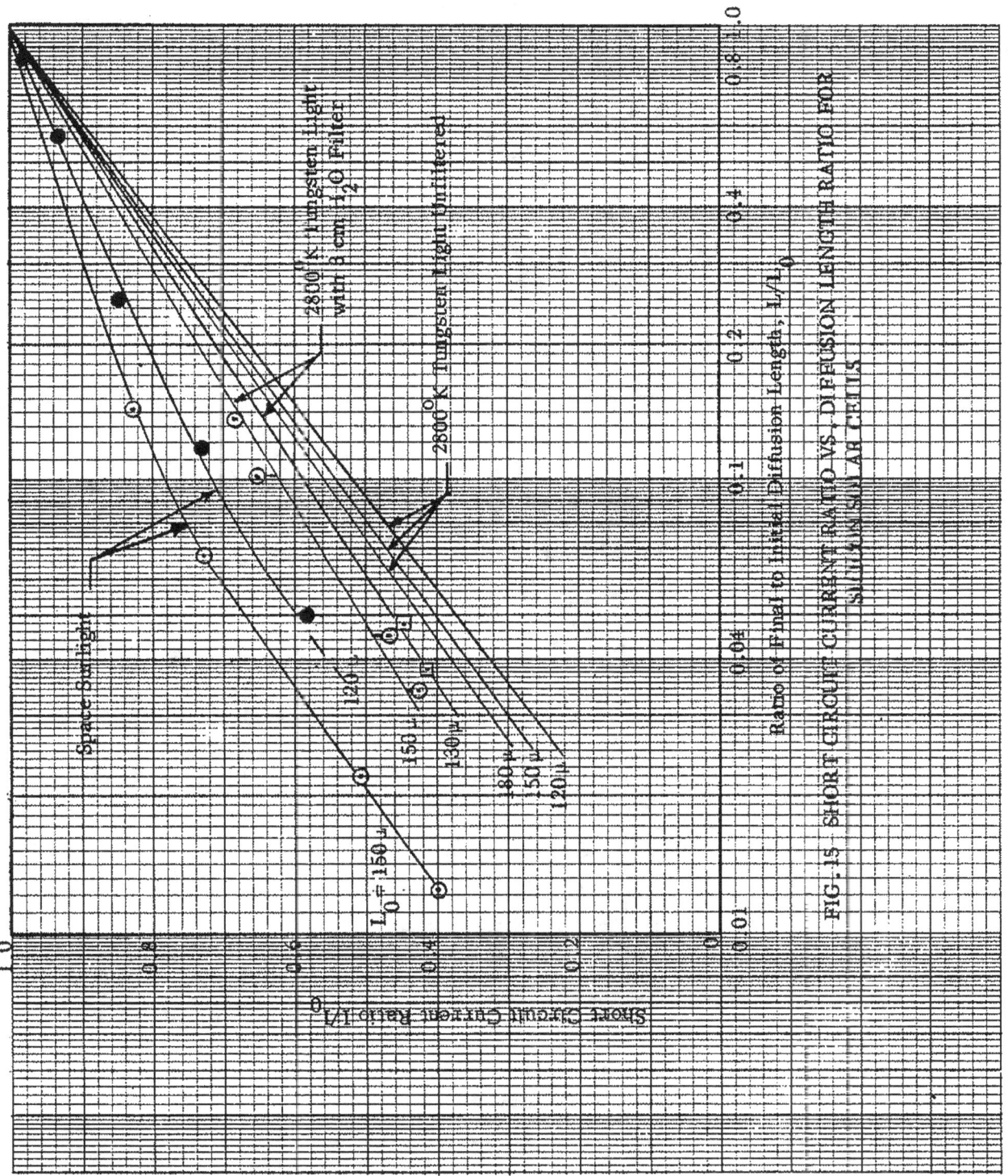

FIG. 15. SHORT CIRCUIT CURRENT RATIO VS. DIFFUSION LENGTH RATIO FOR SILICON SOLAR CELLS

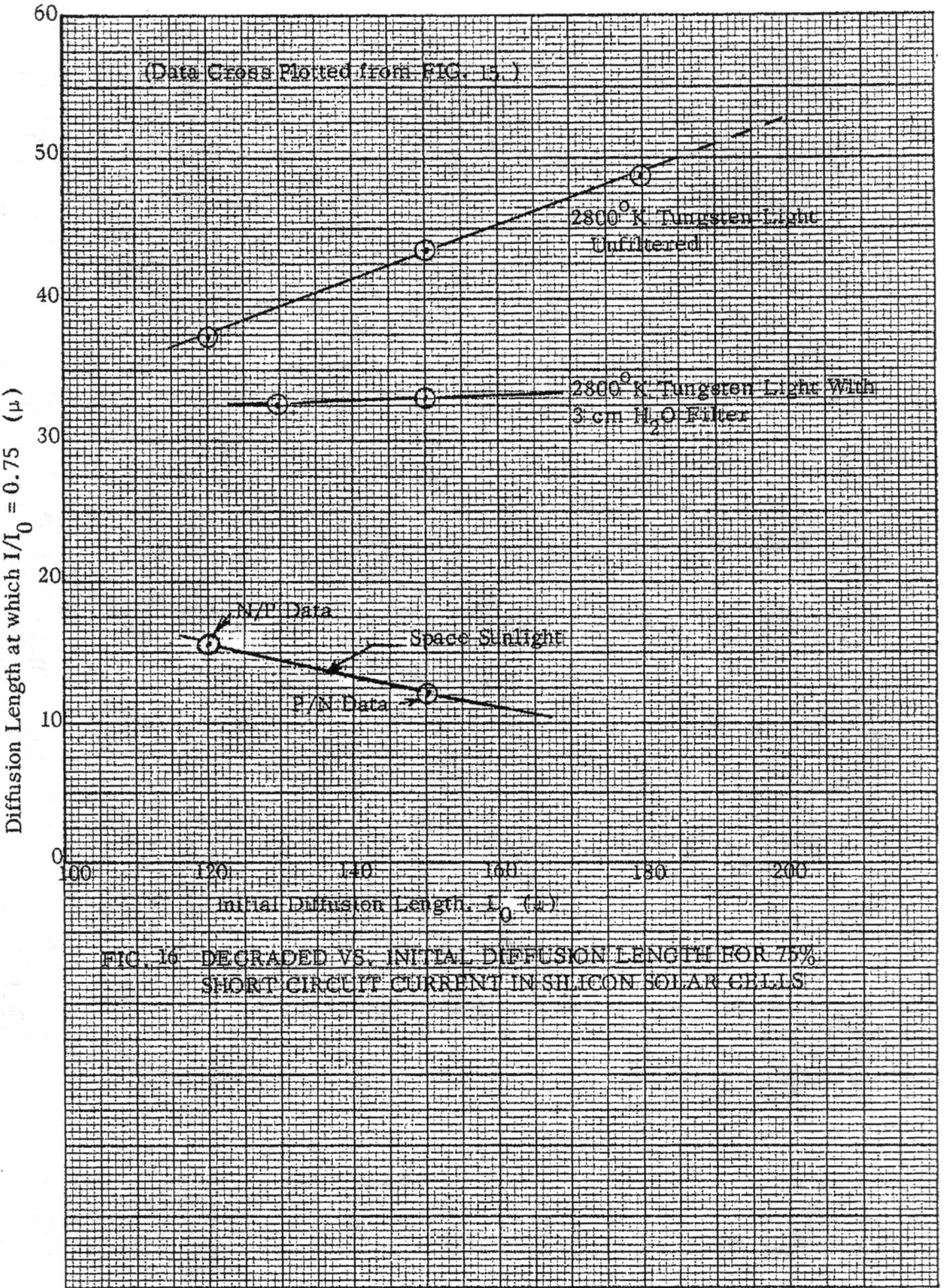

FIG. 16 DEGRADED VS. INITIAL DIFFUSION LENGTH FOR 75% SHORT CIRCUIT CURRENT IN SILICON SOLAR CELLS

computed. This value is calculated to be 4.1×10^5 cm^{-2} based on data for blue n/p cells with $L_0 = 120\mu$ in space sunlight.

4. Maximum Power Output

A typical plot of the variation in the maximum power ratio as a function of the short circuit current ratio is shown in Fig. 17. It is seen that the maximum power decreases only slightly more rapidly than the short circuit current.

Fig. 18 shows the power output of n/p solar cells as a function of cell voltage at various degrees of degradation produced by 1 Mev electrons. These data for space sunlight have been correlated in dimensionless form as maximum power ratio vs. diffusion length ratio in Fig. 19, along with the comparable data for filtered and unfiltered 2800°K tungsten light. Data computed for space sunlight (Ref. 71) for blue p/n cells with an initial diffusion length of 150 μ are included.

5. Power at a Fixed Voltage

The rate of degradation in solar cell current when operated at a fixed cell voltage (as when connected to charge a battery) depends on the value of the operating voltage. If the operating voltage is chosen too high, the cell current will degrade very rapidly. On the other hand, if the cell voltage is too low, then the maximum power available may not be obtained. It is seen from Fig. 18 that if the cell is to operate until its diffusion length is degraded to about 7.6 microns, the cell voltage should be selected at approximately 0.40 volts in order to obtain maximum available power at the end of life. However, consideration of temperature coefficients may force selection of a lower voltage.

6. Temperature Coefficients of Solar Cells as a Function of Radiation Damage

Martin, Teener and Ralph (Ref. 55) have investigated the temperature coefficient of silicon solar cells using a Spectrosun solar simulator. They found that the temperature coefficient of short circuit current for n/p 10 ohm-cm cells increased from an initial value of about 60 μamp/°C to 102.5 μamp/°C after irradiation with a flux of 10^{15} 1 Mev electrons/cm^2. The value for p/n 1 ohm-cm cells increased from 62.5 to 100 μA/°C. However, the temperature coefficient of open circuit voltage stayed constant at about -2.24 mv/°C for the n/p cells and -2.05 mv/°C for the p/n cells. Ross (Ref. 73) measured a temperature coefficient for the maximum power voltage of about -2.08 mV/°C for both n/p and p/n cells.

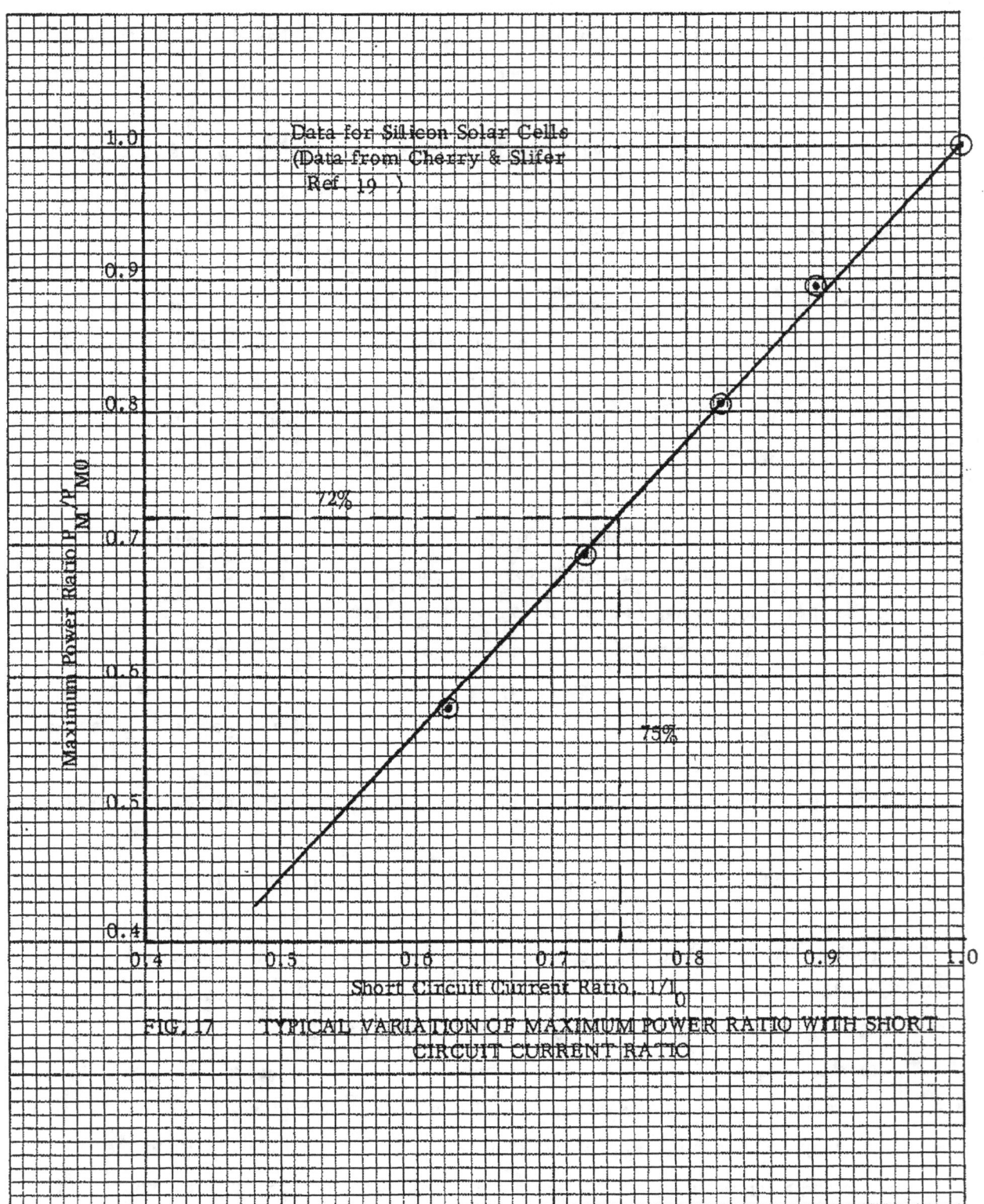

FIG. 17 TYPICAL VARIATION OF MAXIMUM POWER RATIO WITH SHORT CIRCUIT CURRENT RATIO

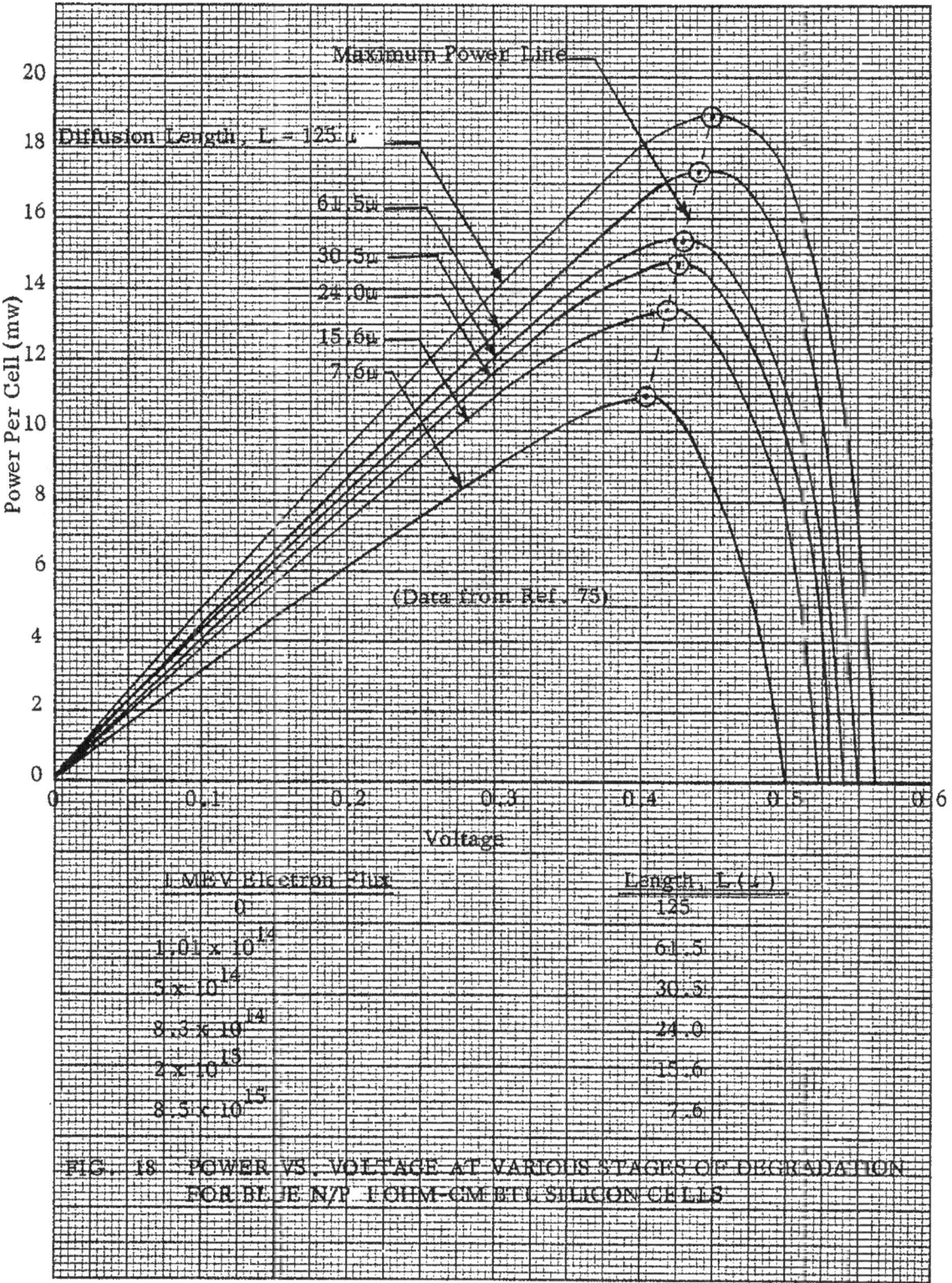

FIG. 18 POWER VS. VOLTAGE AT VARIOUS STAGES OF DEGRADATION FOR BLUE N/P 1 OHM-CM BTL SILICON CELLS

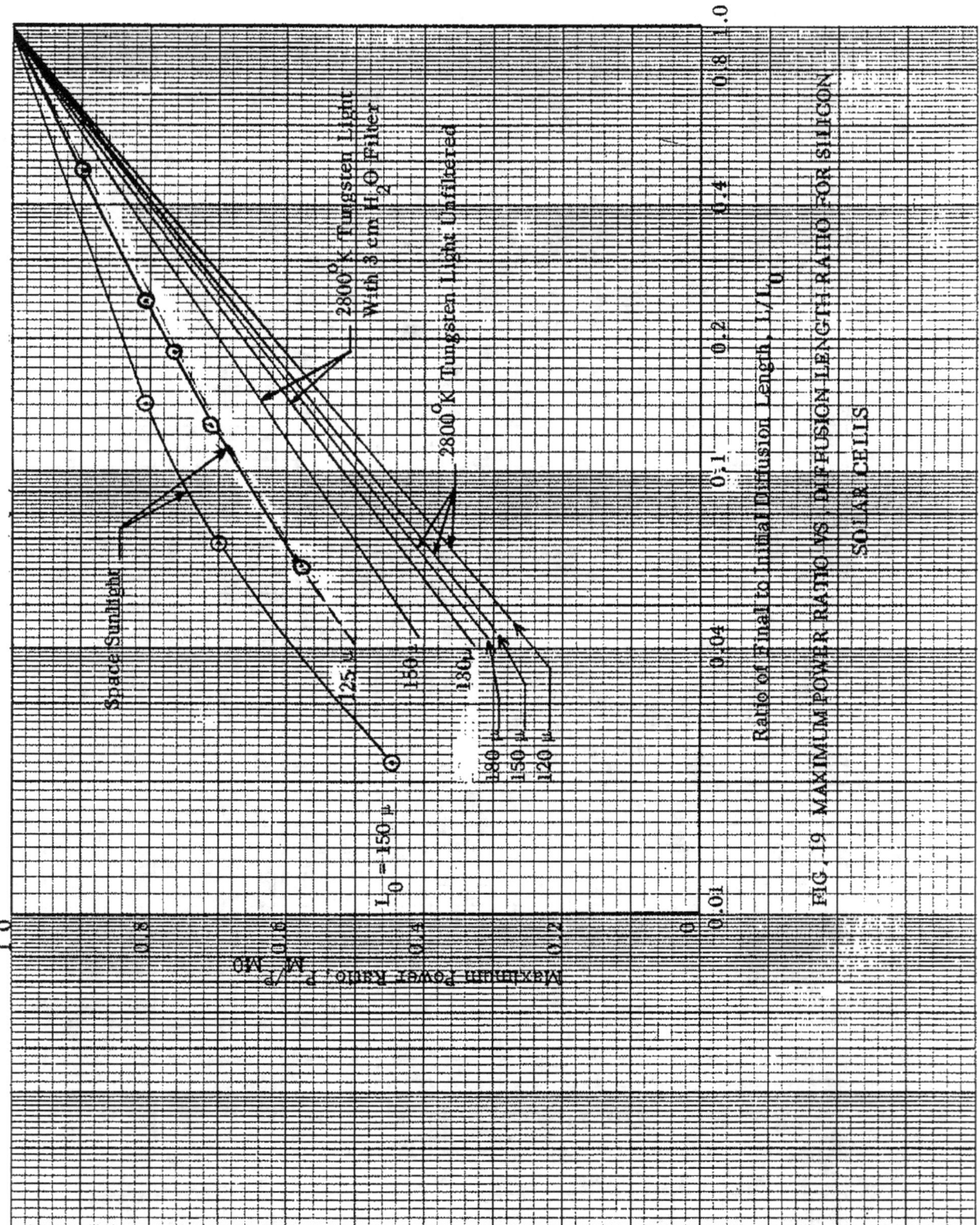

FIG. 19 MAXIMUM POWER RATIO VS. DIFFUSION LENGTH RATIO FOR SILICON SOLAR CELLS

It is important to recognize that the temperature coefficient of short circuit current would be about a factor of ten too high if measured with a tungsten light source, which overemphasizes the red response as compared to sunlight and therefore overstresses the effect of variation of minority carrier lifetime with temperature.

Unless the negative temperature coefficient of open circuit voltage is taken into account, the operating voltage point may be selected too close to the open circuit voltage and very severe degradation of performance will occur as radiation damage proceeds.

Experimental results on the combined effects of temperature and radiation damage on the I-V characteristics of solar cells as reported by Martin, Teener and Ralph (Ref. 55) are shown in Fig. 20, 21, 22, and 23. By reference to these figures, the operating voltage point may be selected so that maximum power is obtained after degradation to the required value of equivalent 1 Mev electron flux (or corresponding diffusion length).

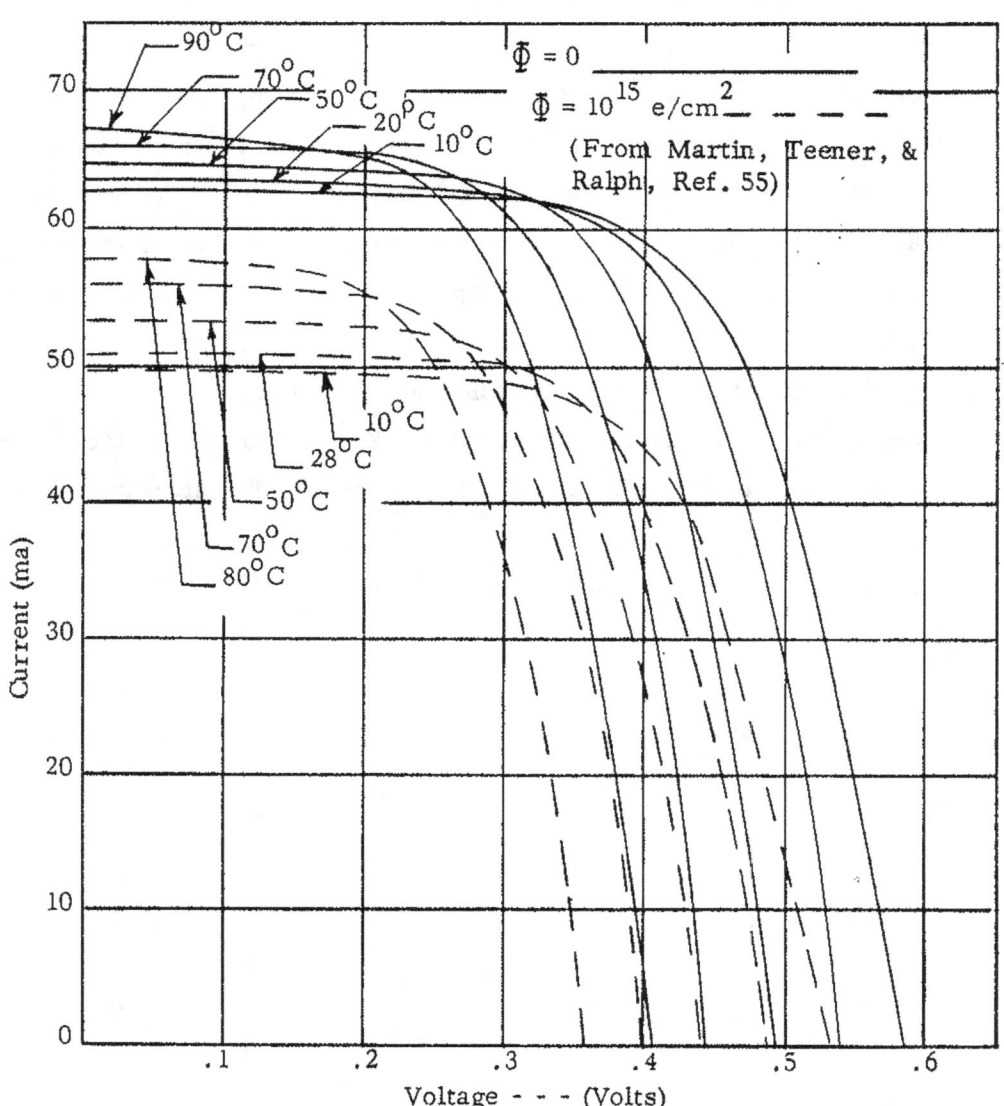

FIG. 20 CURRENT VS. VOLTAGE CURVES FOR N/P 10 OHM-CM SILICON SOLAR CELLS AT VARIOUS CELL TEMPERATURES BEFORE AND AFTER 1 MEV ELECTRON IRRADIATION

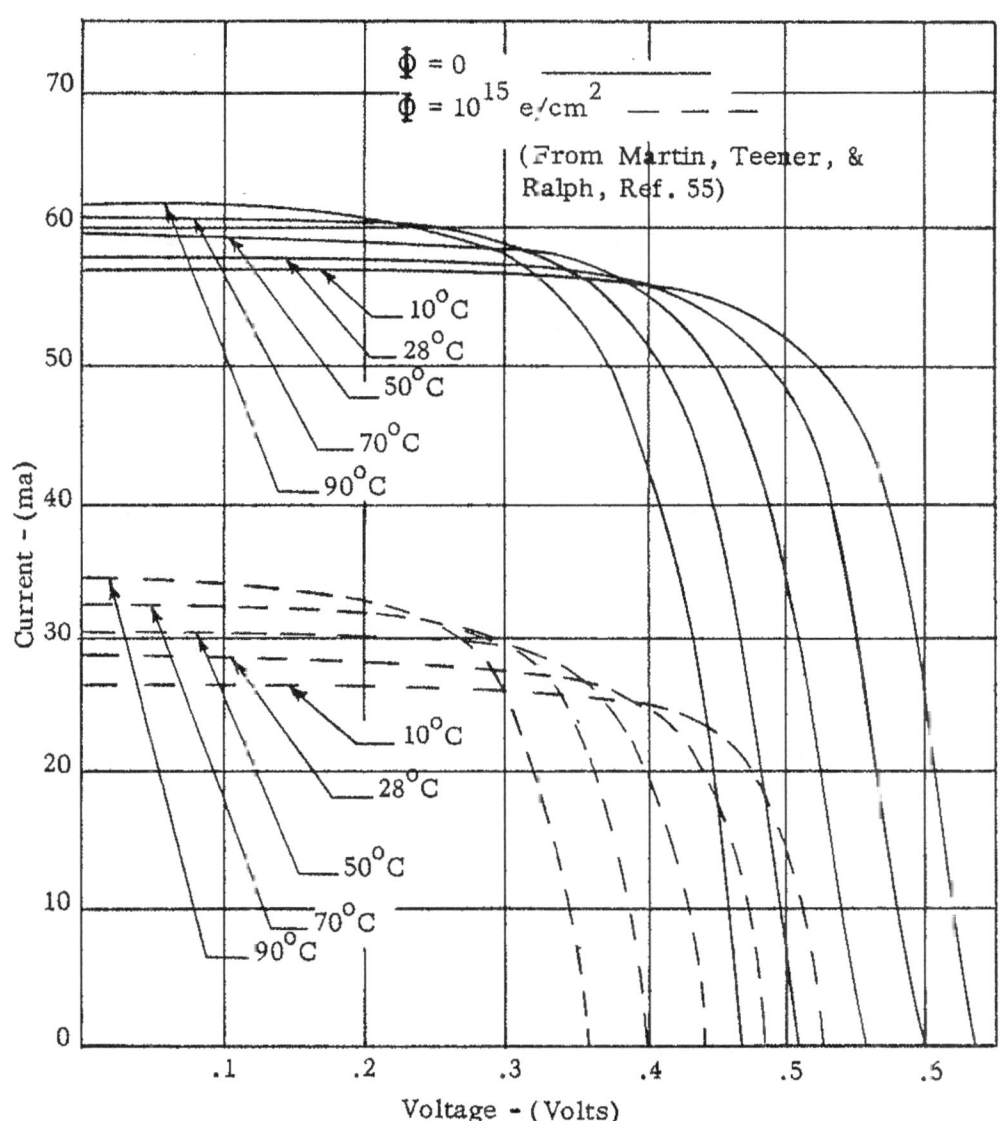

FIG. 21 CURRENT VS. VOLTAGE CURVES FOR P/N 1 OHM-CM SILICON SOLAR CELLS AT VARIOUS CELL TEMPERATURES BEFORE AND AFTER 1 MEV ELECTRON IRRADIATION

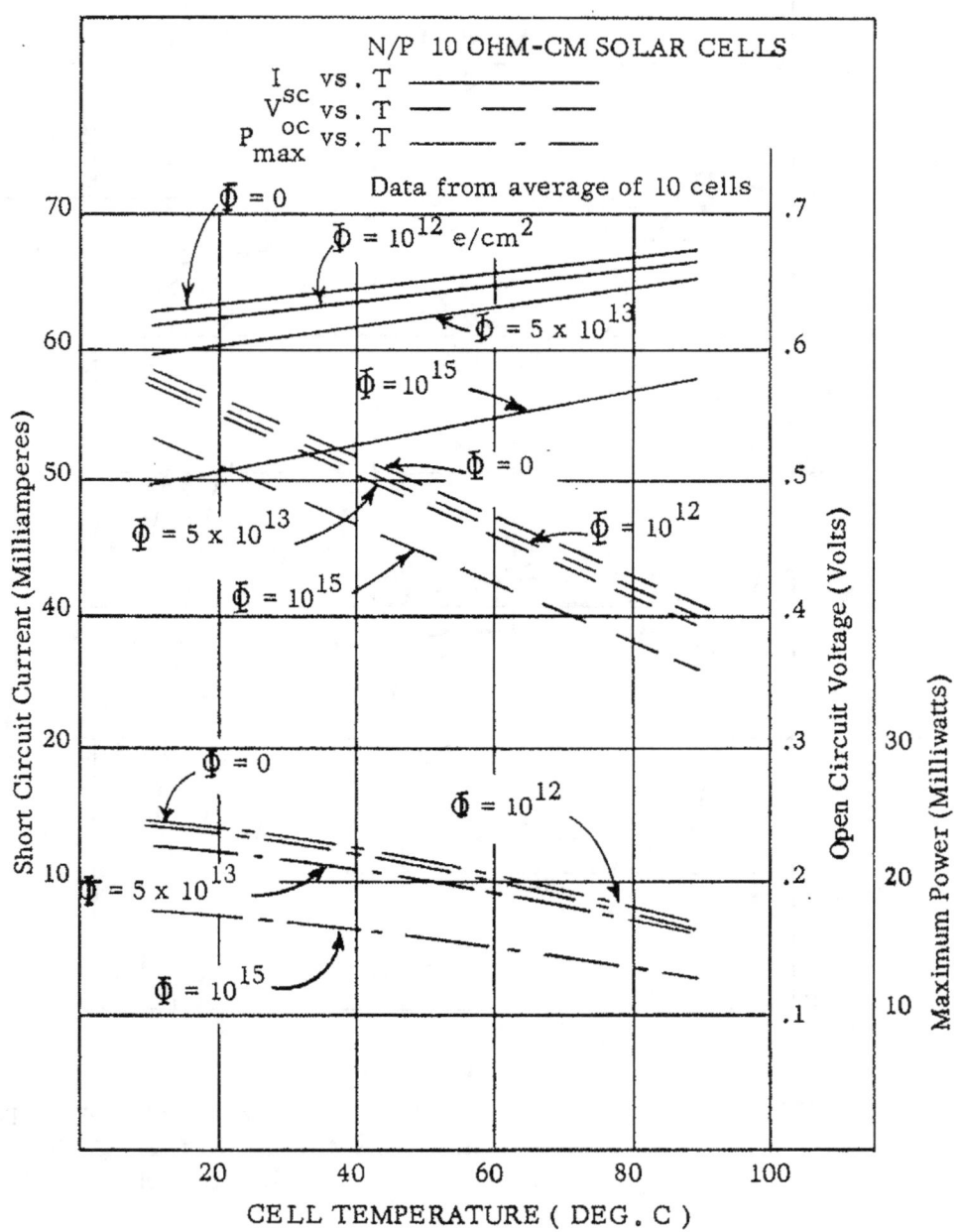

FIG. 22 ELECTRICAL CHARACTERISTICS AS FUNCTIONS OF TEMPERATURE AT INCREASING LEVELS OF 1 MEV ELECTRON FLUX FOR N/P SOLAR CELLS

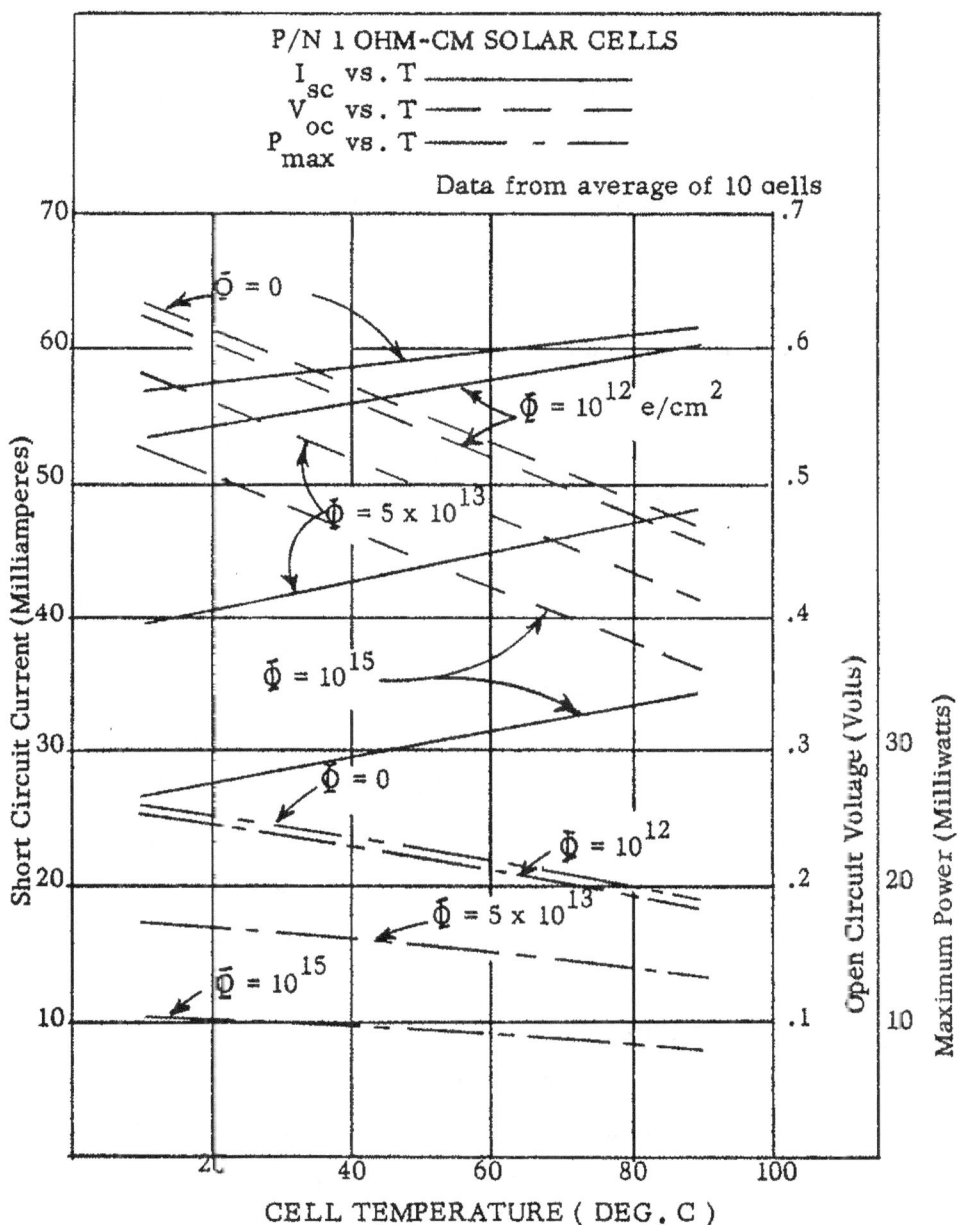

FIG. 23 ELECTRICAL CHARACTERISTICS AS FUNCTIONS OF TEMPERATURE AT INCREASING LEVELS OF 1 MEV ELECTRON FLUX FOR P/N SOLAR CELLS

IV THE SPACE RADIATION ENVIRONMENT
(See also recent Reference 145 in the July 1963 issue of the Bell System Technical Journal)

A. Introduction

The most damaging space radiations for solar cells are the geomagnetically trapped electrons and protons in the magnetosphere and the solar flare protons which occur sporadically at times near maximum solar activity and which are observed primarily outside the magnetosphere. Other radiations which are generally negligible with respect to radiation damage on semiconductors include primary galactic cosmic rays (which are primarily protons and heavier ions), neutrons produced by cosmic ray interaction with the atmosphere, solar x-rays and gamma rays.

For earth satellites within the magnetosphere, it is important to know the flux and energy spectra of electrons and protons encountered in order to calculate radiation effects. An excellent survey article as of October 1962 was prepared by O'Brien (Ref. 104). Unfortunately, recent experimental data on trapped electrons and protons has not yet been properly correlated to provide a good knowledge of the flux and energy spectrum of trapped particles at all locations. Furthermore the "Starfish" nuclear explosion of July 9, 1962 injected large numbers of artificial electrons which are still decaying so that the electron fluxes are still changing with time. Therefore, the brief summary below must be considered tentative and subject to change after further correlation of experimental measurements.

B. Mc Ilwain's Coordinates

In order to discuss the distribution of trapped particles it is most convenient to use the coordinate system developed by Mc Ilwain (Ref. 99). The two coordinates are B, the scalar magnitude of the magnetic field, and L, a parameter that is very nearly constant along each magnetic field line and which is nearly equal to the radial distance to the field line at the geomagnetic equator. Fig. 24 shows an example of L lines plotted with respect to magnetic dipole coordinates R and λ, taken from (Ref. 99) where R is the distance from the center of the magnetic dipole and λ is the geomagnetic latitude.

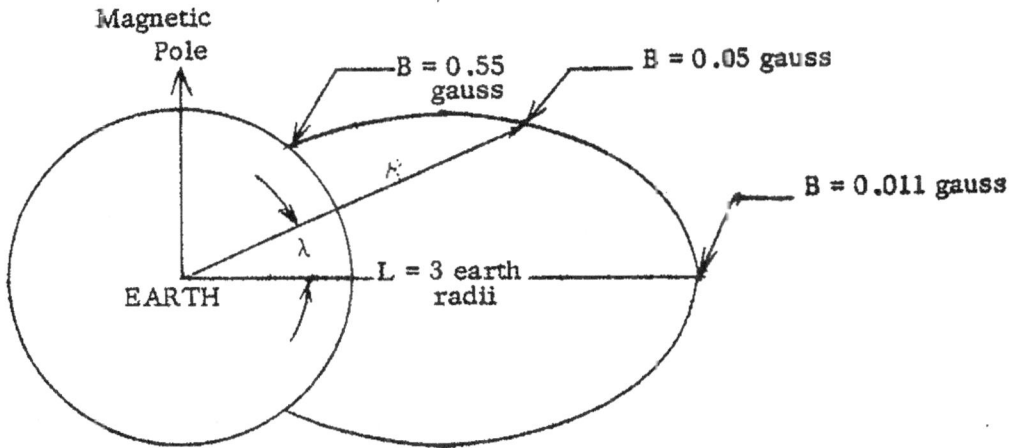

FIG. 24 THE B-L MAGNETIC COORDINATE SYSTEM

The trapped particles may be visualized as moving primarily in helical paths around each magnetic field line (~ constant L) and being reflected between mirror points north and south of the magnetic equator at points where the field B has increased over its value at the equator by the ratio ($1/\sin^2 \alpha_o$) where α_o is pitch angle of particle at magnetic equator. Each particle bounces between the north and south mirror points with a period of the order of seconds or less. In addition, electrons drift eastward and protons westward with a period of the order of a minute required to circle the earth, although this longitudinal motion can be neglected in understanding the steady state flux distributions in the belt.

The fluxes and energy spectra of particles can be correlated analytically quite well with respect to the B-L coordinate system. The pitch angle distribution of the particles can also be correlated with the variation in the magnitude of the field B along each line of constant L. The flux is higher perpendicular to the magnetic field line than parallel to the field line, particularly at the mirroring points. This effect of pitch angle distribution is significant for a satellite (like ANNA-1B) which is oriented with respect to the magnetic field. However, generally the flux may be defined in terms of an average omnidirectional flux at a point in B-L space.

An alternate plot of the relationship between the B-L coordinates and the geomagnetic coordinates R and λ is shown in Fig. 25. In converting from B-L coordinates to earth-centered coordinates it must be recognized that the earth's magnetic field, if approximated as a simple dipole, is centered at a point located 411.4 Km from the center of the earth on a line toward the point at 150.8 degrees east longitude and 15.6 degrees north latitude (Ref. 3). The axis of the dipole is tilted at an angle of 11.7 degrees from the earth's axis toward the geographic longitude of 69 degrees west.

Even if fluxes in the B-L coordinates were known precisely, in order to predict fluxes at an arbitrary point in space with an accuracy better than a factor of 10 it is necessary to include the anomalous variations in the magnetic field of the earth from the dipole approximation. An important example is the South Atlantic Anomaly, where magnetic field lines dip to low altitudes and appreciable fluxes of trapped protons and electrons are encountered at altitudes of only 1000 Km.

C. Electron Fluxes and Energy Spectra

The flux of natural electrons in the inner region of the magnetosphere have been augmented by artificial electrons injected by high altitude nuclear explosions. Table IV-1 lists the explosions which have produced measurable effects. Of these, only the "Starfish" explosion on July 9, 1962 has produced a long-lasting belt of electrons which is of importance in affecting solar cells.

Hess (Ref. 96) has prepared maps of the artificial electron belt in B-L and R-λ coordinates (Fig. 26 and 27) which are applicable for a time of one week after the "Starfish" explosion on the assumption that the electrons all have a fission spectrum as shown in Fig. 28. This map by Hess showed appreciable fluxes of electrons at values of L up to 4 (altitudes up to 3 earth radii at the magnetic equator). However, data by Van Allen (Ref. 111) indicated that the fission electrons were confined to lower values of L. It now appears that the fission spectrum is approximately applicable only for L less than 2 and that the electron spectrum is softer (a steeper variation of flux with energy) than a fission spectrum for L greater than 2, where many of the electrons observed were naturally present prior to "Starfish". An effort is now being devoted by Hess at the Goddard Space Flight Center to correlate the electron spectra as a function of L (Ref. 40). Until this correlation becomes available it seems appropriate to assume

Plotted according to the transformation:

$$B = \frac{M}{R^3} \left(4 - \frac{3R}{L}\right)^{1/2} \qquad R = L \cos^2 \lambda$$

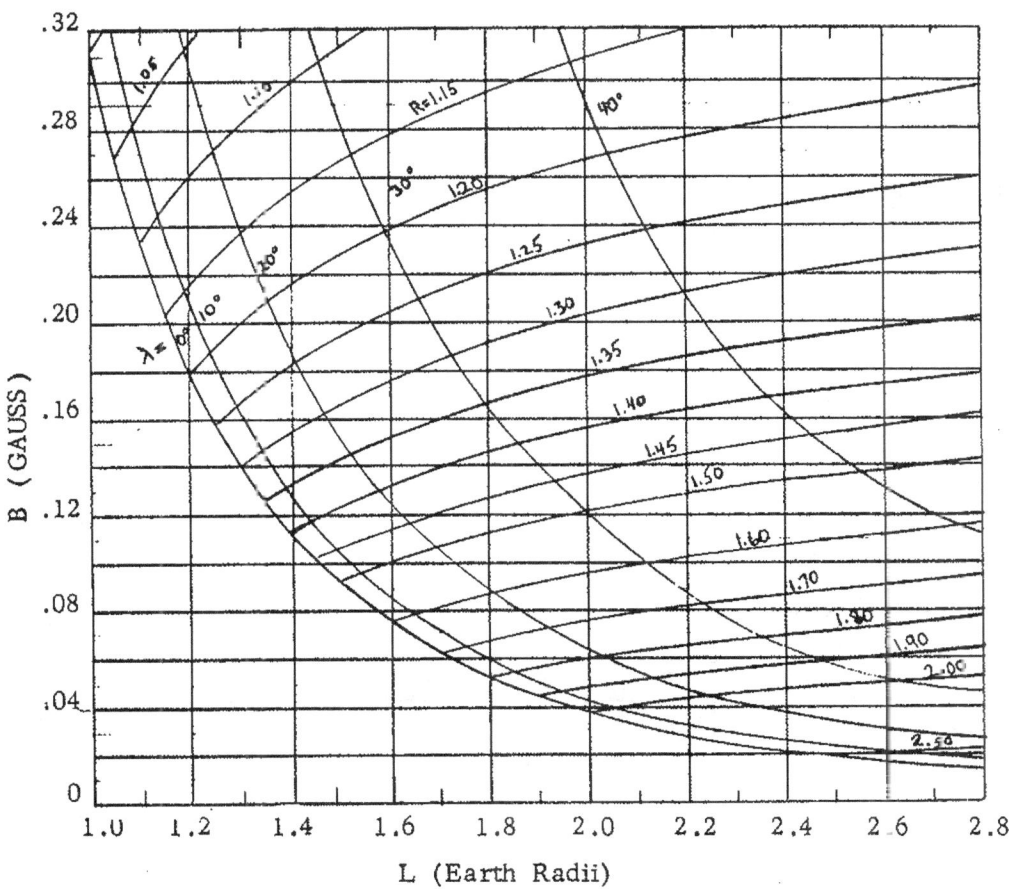

FIG. 25 THE MAPPING OF THE POLAR COORDINATES R and λ ONTO THE B - L PLANE

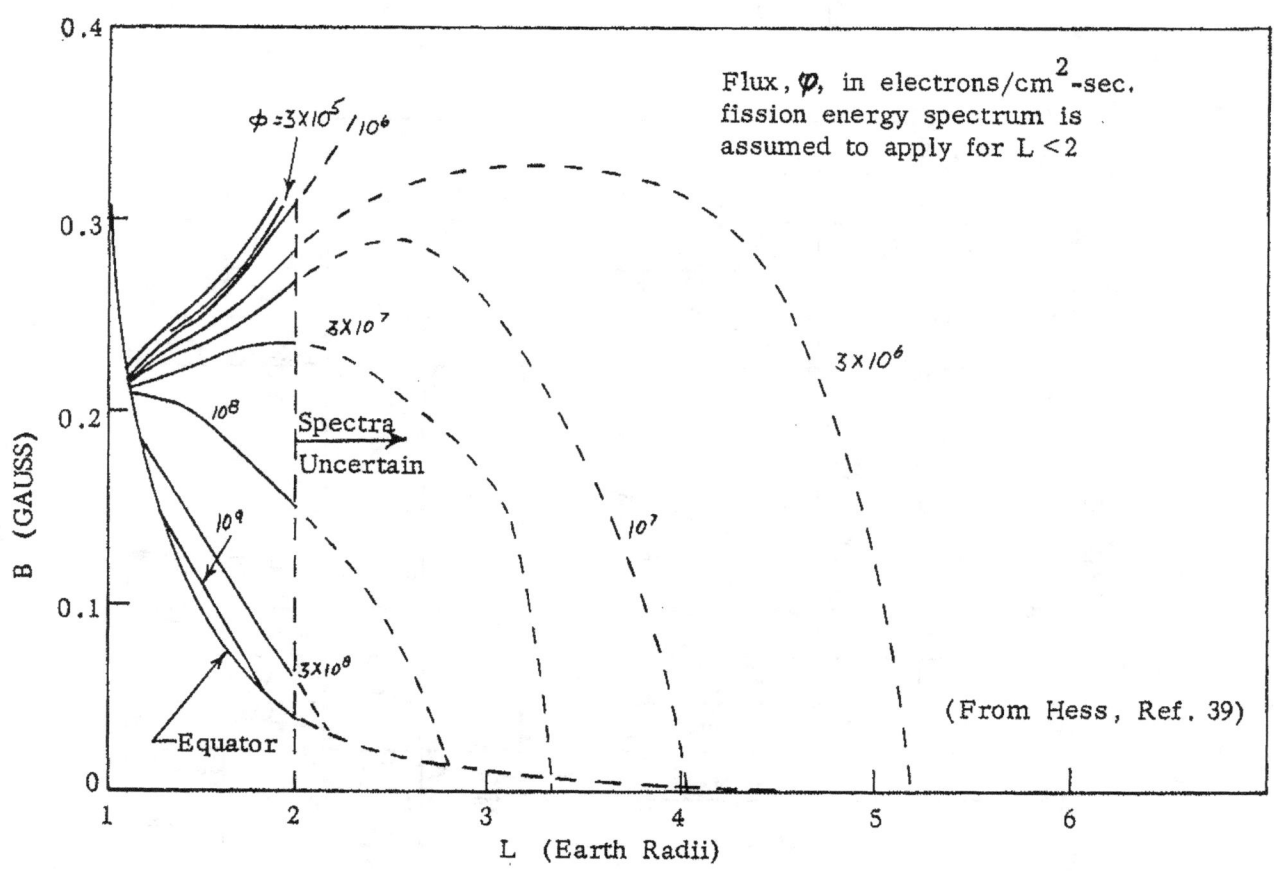

FIG. 26 THE B - L MAP OF ELECTRON FLUXES APPROXIMATELY ONE WEEK AFTER STARFISH

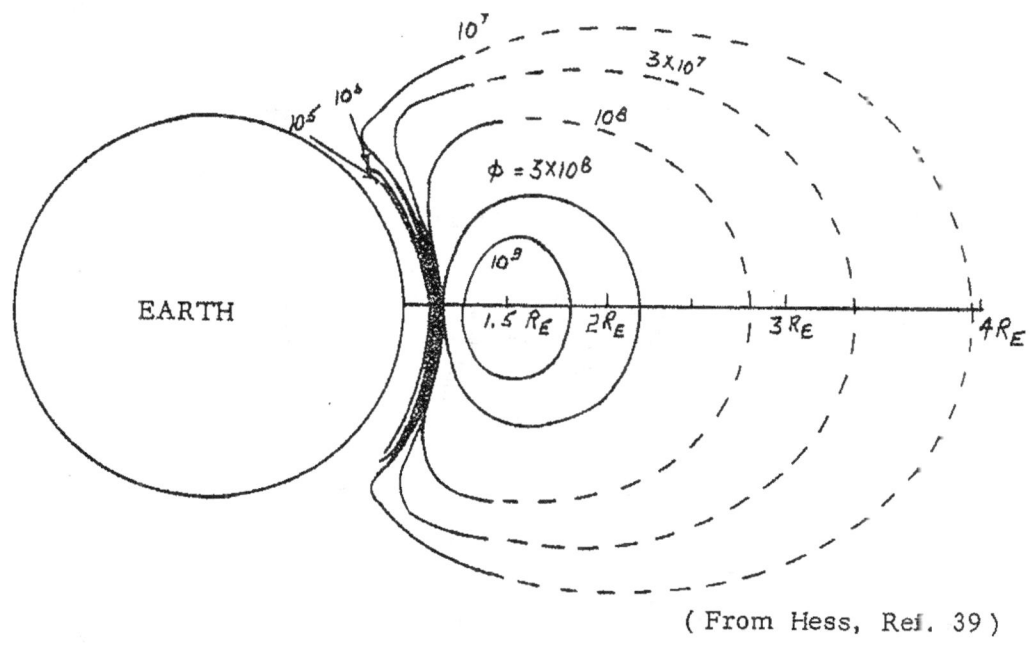

(From Hess, Ref. 39)

FIG. 27 THE R - λ MAP OF ELECTRON FLUXES APPROXIMATELY ONE WEEK AFTER STARFISH

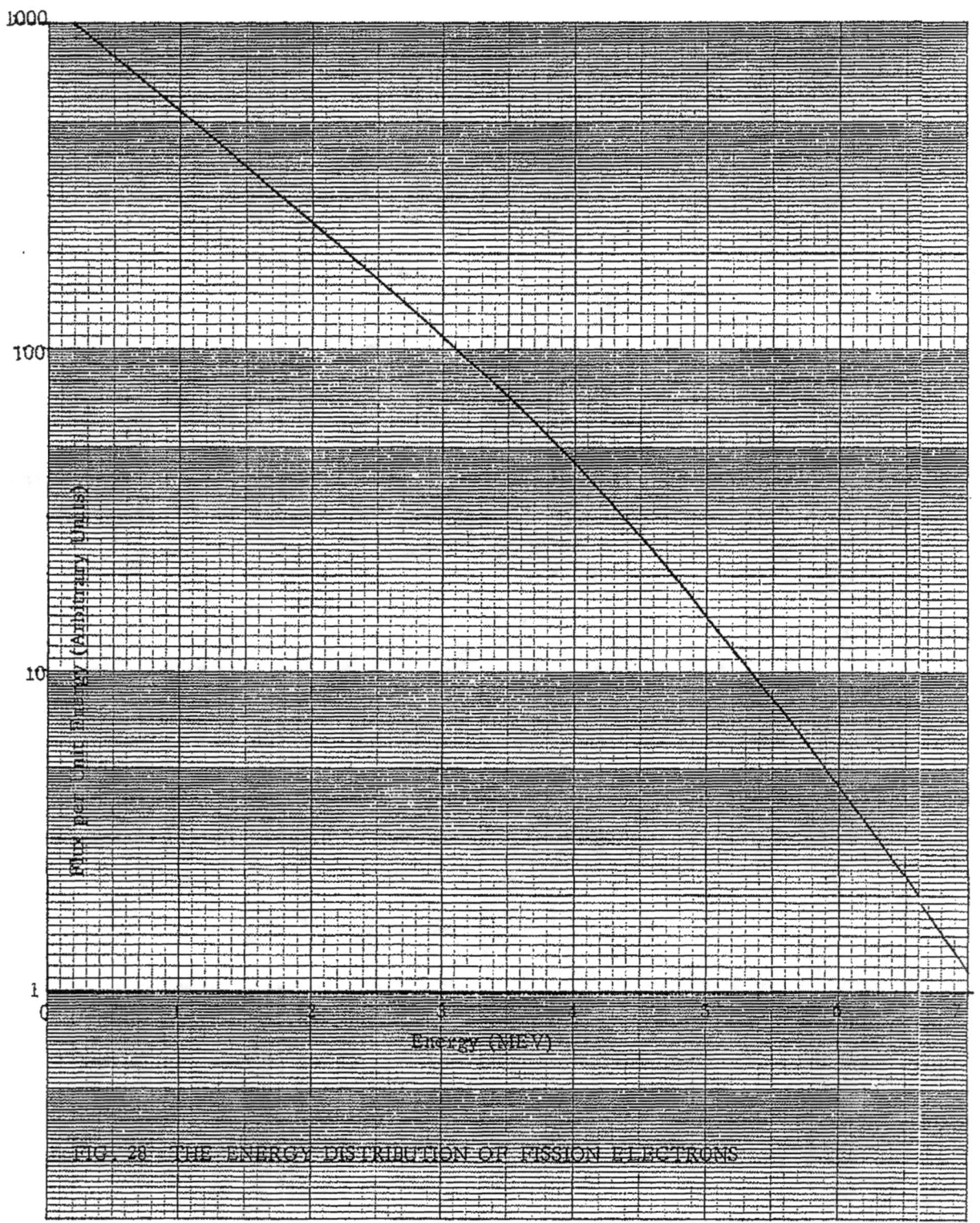

FIG. 28 THE ENERGY DISTRIBUTION OF FISSION ELECTRONS

that Hess' map as shown in Fig. 27 applies approximately for values of L less than 2 and that the fission spectrum as shown in Fig. 28 applies in this region. However, it is clear that the electron fluxes have continued to decay since one week after Starfish and that the flux values appropriate at a time of 5-1/2 months after Starfish (Dec. 1962) may be approximately a factor of two lower than shown in Fig. 27. Until further correlation work is completed, it does not appear useful to try to calculate electron fluxes and spectra for L values from 2 to 4.

The average electron fluxes in the outer zone (i.e., for values of L greater than 4) are shown in Fig. 29 (Ref. 93). It should be noted that the electron flux in this region fluctuates radically during each day (Ref. 103) and also as a result of geomagnetic storms (Ref. 108) and that the applicable energy spectra in this region are not well established. It is seen in Fig. 29 that the average shape of the constant flux contours in the magnetic equatorial plane are distorted by interaction with plasma from the sun (the solar wind).

D. Proton Fluxes and Energy Spectra

McIlwain and Pizzella (Ref. 100) have shown that the energy spectrum of protons throughout the magnetosphere may be approximated by the equation:

$$\rho_p(E)\,dE = (\text{Constant})\ e^{-E/E_o}dE \tag{9}$$

with

$$E_o = 306\,L^{-5.2}\ \text{Mev} \tag{10}$$

This equation gives a reasonable fit to the data for values of L up to 8 and is in not bad agreement with Naugle and Kniffen's spectra (Ref. 102) obtained with an emulsion experiment at L values from 1.47 to 1.79.

The value of the constant which defines the absolute flux at any point may be approximated by using the proton flux map of Fig. 30 (Ref. 101), which shows contours of the integral flux above 31 Mev. Improved proton maps are being prepared by Walter L. Brown of Bell Telephone Laboratories and by the Goddard Space Flight Center in cooperation with the experimenters.

Additional pertinent references on the space environment are included on page R-7. It is expected that improved correlations of data on the fluxes and energy spectra of trapped radiation will become available in the near future, in which case more accurate calculation of space radiation effects will become feasible.

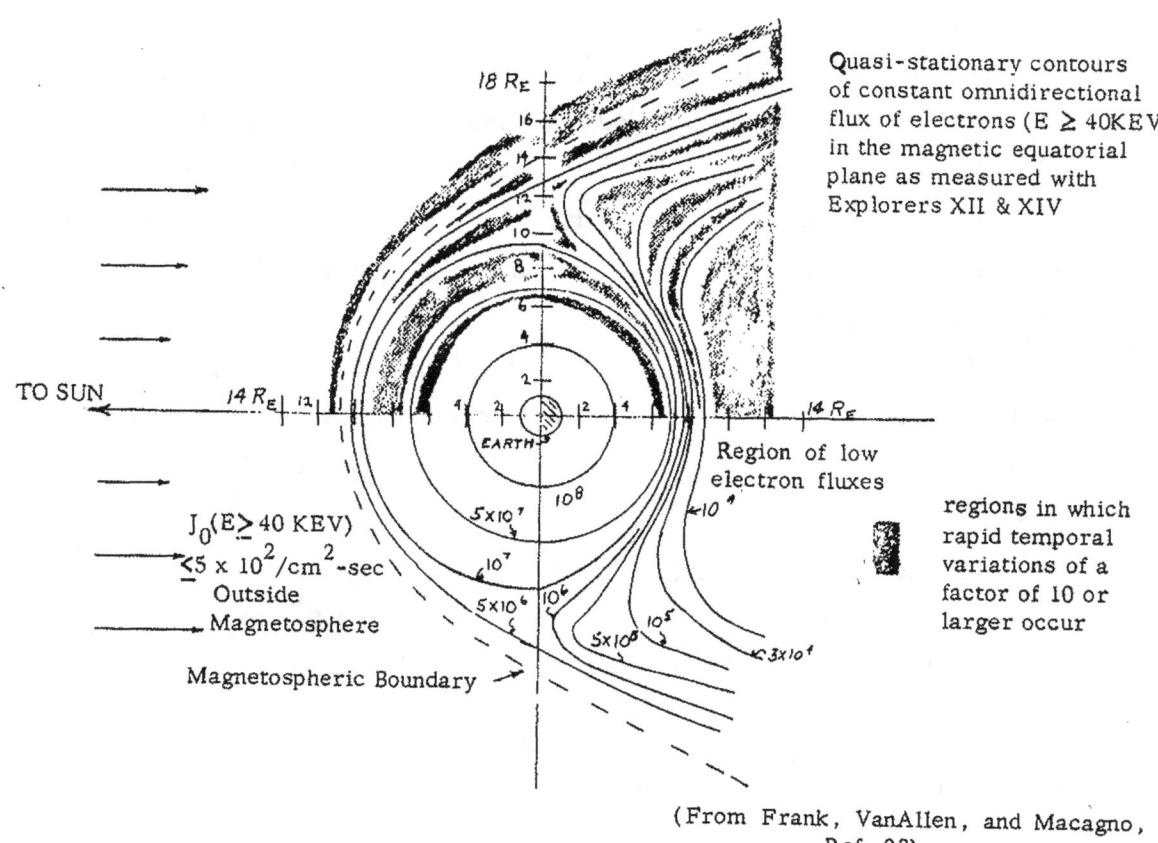

FIG 29. Summary of observed omnidirectional intensities of electrons ($E \gtrsim 40$ kev) obtained from approximately twenty complete orbits of Explorer 14 during October-December 1962 and from Explorer 12 during August-December 1961.

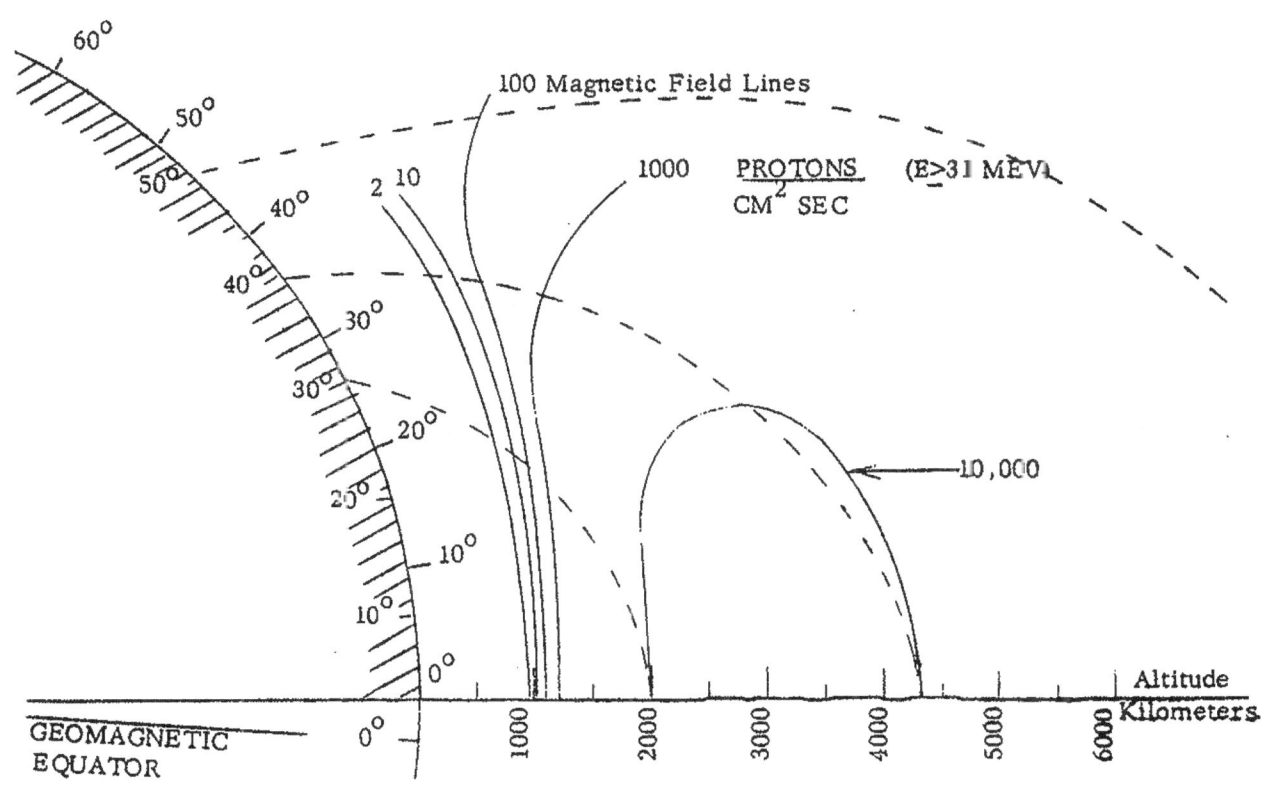

FIG. 30 PROTON FLUX MAP

TABLE IV-1

Data On Nuclear Explosions Which Injected Particles Into The Magnetosphere

Country	Identification	Size	Location	Date	Remarks
U.S.	Teak & Orange	1 MT	Less than 50 mi. above Johnston Island	Aug 1 and 12, 1958	Effects disappeared in a few days due to particle absorption in the South Atlantic Anomaly.
U.S.	Argus (3 shots)	1-2 KT	300 mi. above South Atlantic	1958	Artificial belts lasted several weeks.
U.S.	Starfish	1.4 MT	250 mi. above Johnston Island	7/9/62	Produced long-lasting belt of artificial electrons.
USSR	3 shots	?	High altitude and high latitude	10/22/62 10/28/62 11/1/62	Artificial electron belts in slot between inner and outer zones decayed rapidly.

V CORRELATION OF SATELLITE TEST DATA ON SOLAR CELL PERFORMANCE

This section summarizes the results of telemetered performance data for solar cells with various thicknesses of transparent shielding on satellites which have been subjected to radiation in the natural and artificial radiation belts.

Table V-1 shows the orbital parameters and date of launch of fourteen satellites for which telemetered solar cell data were available. In most cases, the telemetered data were available in the form of short circuit current with normal solar incidence angle as a function of time in orbit. Although the temperature of the solar cells was telemetered in some cases, no effort has been made to correct all data to a standard temperature.

An example of the variation of short circuit current ratio (I/I_0) with time for p/n and n/p silicon solar cells with various shield thicknesses on three satellites is shown in Fig. 31. The flight test data were extrapolated where necessary to estimate the time in orbit required for the short circuit current to decrease by 25%. These extrapolations were made in most cases by assuming a 20% reduction in short circuit current per decade of time after at least 10% degradation had occurred, except where test data showed a different trend. This extrapolation procedure is inaccurate when the radiation exposure is not constant with time, which is the case for many satellites where apogee precession affects the flux encountered per orbit. Table V-2 shows the data on the extrapolated life to 75% of initial short circuit current for solar cells on various satellites. The shield thicknesses have been converted to grams per square centimeter.

The estimated time in days for the n/p solar cells to degrade by 25% have been plotted against the shield thickness in grams per square centimeter in Fig. 32. A similar plot of the data for p/n solar cells is shown in Fig. 33.

An inspection of the data showed that for the Relay-1 and Anna 1-B satellites, which carried both n/p and p/n cells, there was a factor in the neighborhood of four times longer life for n/p cells than for p/n cells under shield thicknesses greater than

TABLE V-1

ORBITAL PARAMETERS OF SATELLITES

No.	Satellite	Perigee	Apogee	Inclination	Period (Min.)	Launch Date
1.	Explorer XI	490.8 KM 264 N.MI.	1799 KM 970 N.MI.	28.8°		Apr. 27, 1961
2.	Midas III	3450 KM 1850 N.MI.	3510 KM 1890 N.MI.	91.2°	161.5	July 12, 1961
3.	Explorer XII	304 KM 163 N.MI.	77,000 KM 41,500 N.MI.	33°		Aug. 16, 1961
4.	Midas - IV	3530 KM 1898 N.MI.	3760 KM 2025 N.MI.	95.9°	166	Oct. 21, 1961
5.	TRAAC	960 KM 516 N.MI.	1106 KM 596 N.MI.	32°	105.6	Nov. 15, 1961
6.	TRANSIT - IV B	960 KM 516 N.MI.	1106 KM 596 N.MI.	32°	105.6	Nov. 15, 1961
7.	Ariel - I	390 KM 210 N.MI.	1210 KM 650 N.MI.	54°	100.8	Apr. 26, 1962
8.	Telstar - I	952 KM 513 N.MI.	5660 KM 3040 N.MI.	45°	157.7	July 10, 1962
9.	Tetrahedral Research Satellite - I	112 N.MI.	362 N.MI.	82°		Sept. 17, 1962
10.	Alouette (S-27)	1004 KM 540 N.MI. 1.16 R_E	1029 KM 554 N.MI. 1.16 R_E	POLAR	105.5	Sept. 29, 1962
11.	Explorer XIV	278 KM 150 N.MI.	99,000 KM 53,000 N.MI.	33°	2185 (36.4 hr.)	Oct. 21, 1962
12.	1962 - βK (STARAD)	191 KM 103 N.MI. 1.03 R_E	5550 KM 3000 N.MI. 1.87 R_E	71°		Oct. 26, 1962
13.	ANNA - 1B	1090 KM 584 N.MI. 1.17 R_E	1180 KM 635 N.MI. 1.19 R_E	50°		Oct. 31, 1962
14.	Relay - I	1321 KM 713 N.MI. 1.21 R_E	7439 KM 4001 N.MI. 2.17 R_E	47.5°	185.1	Dec. 13, 1962

TABLE V-2

Estimated Time in Orbit to Degrade Cells to 75% of Initial Short Circuit Current
(Based On Telemetry Data)

Satellite	Type of Cell	Mils Shielding and Material	Shield Thickness gm/cm^2	$T_{0.75}$ Days	Remarks
1. Explorer-XI (Ref. 47)	n/p	Bare	0	70	Low energy protons.
2. Midas - III (Ref. 66)	p/n gridded	40 Silica	0.275	100	Data corrected to 44° C.
	p/n gridded	80 Silica	0.55	250	
	n/p	60 Silica	0.41	1000	
3. Explorer-XII	p/n	3 Glass	0.02	1000	Low energy protons.
	p/n	Bare	0	1 Orbit	
4. Midas - IV (Ref. 66)	p/n	40 Silica	0.275	55	Data at 87° C.
	p/n	80 Silica	0.55	110	
5. TRAAC (Ref. 34)	p/n	6 Glass	0.038	-	Knocked out by Starfish.
6. TRANSIT - IVB (Ref. 34)	p/n	6 Glass	0.038	-	Knocked out by Starfish.
7. Ariel - I	p/n			-	Knocked out by Starfish. (Partially)
8. Telstar - I (Ref. 16)	n/p	30 Sapphire	0.3	400	
9. Tetrahedral Research Satellite #1 (Ref. 26)	p/n and others	Bare	0	∞	Low alt. orbit.
10. Alouette (S-27)	p/n	12 Glass	0.076	30	
11. Explorer-XIV	p/n Blue	Bare	0	1 Orbit	Low energy protons.
12. 1962 ϑ Κ (Ref. 33)	p/n	6 Glass	0.038	4	Affected by Russian explosions.
	p/n	60 Glass	0.38	100	
	n/p	6 Glass	0.038	50	
13. ANNA - 1B (Ref. 33)	p/n	6 Glass	0.038	40	Damage occurs in South Atlantic Anomaly
	p/n	20 Silica	0.135	90	
	p/n	30 Sapphire	0.3	400	
	n/p	30 Sapphire	0.3	1400	
14. Relay - 1 (Ref. 81 & 82)	p/n	Bare	0	0.04	(Ref. 81)
	p/n	30 Silica	0.2	30	(Ref. 81)
	p/n	60 Silica	0.4	46	(Ref. 81)
	n/p	Bare	0	0.35	(Ref. 81)
	n/p	30 Silica	0.2	160	(Ref. 82)
	n/p	60 Silica	0.4	240	(Ref. 82)

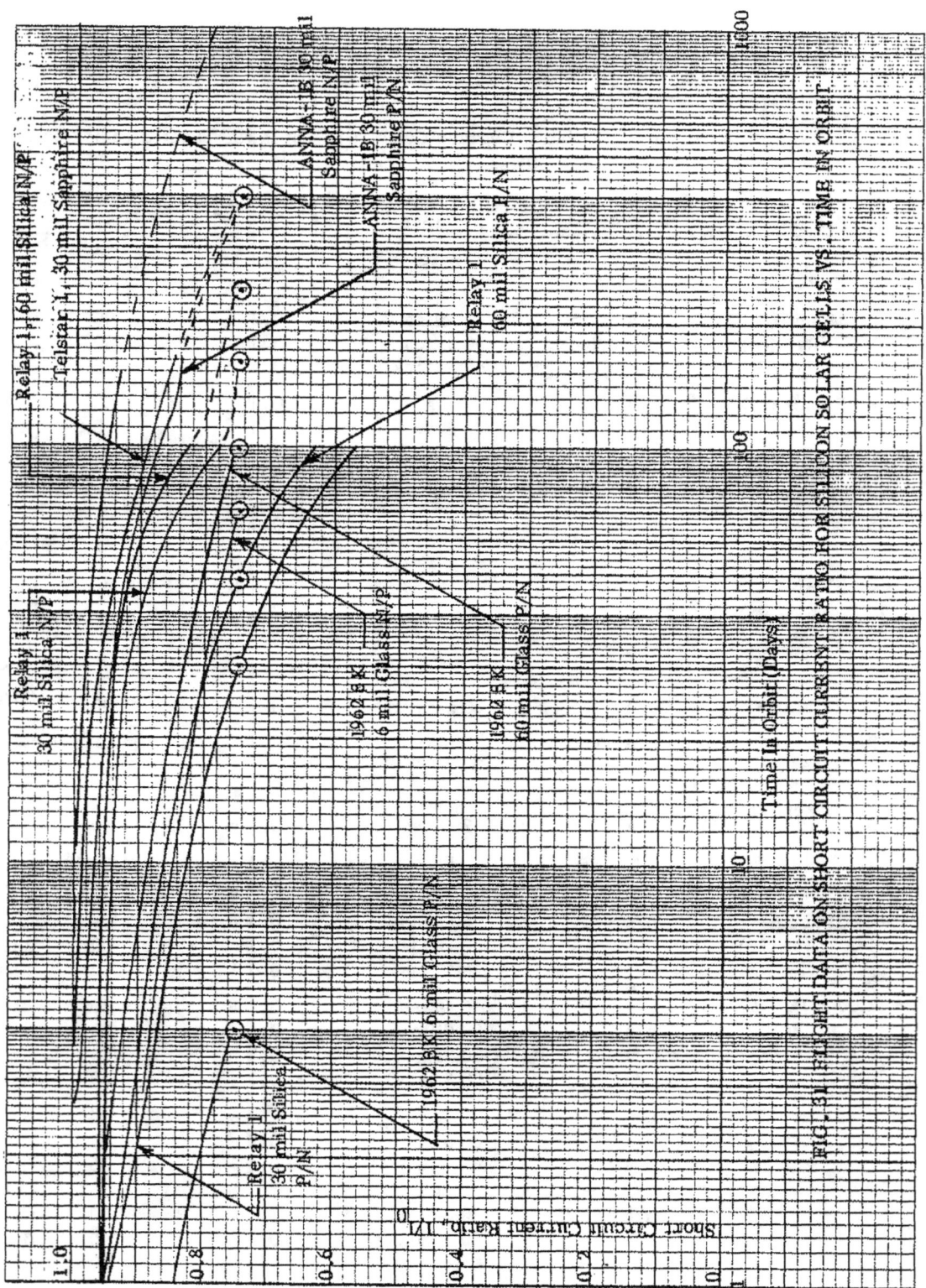

FIG. 31 FLIGHT DATA ON SHORT CIRCUIT CURRENT RATIO FOR SILICON SOLAR CELLS VS. TIME IN ORBIT

- 58 -

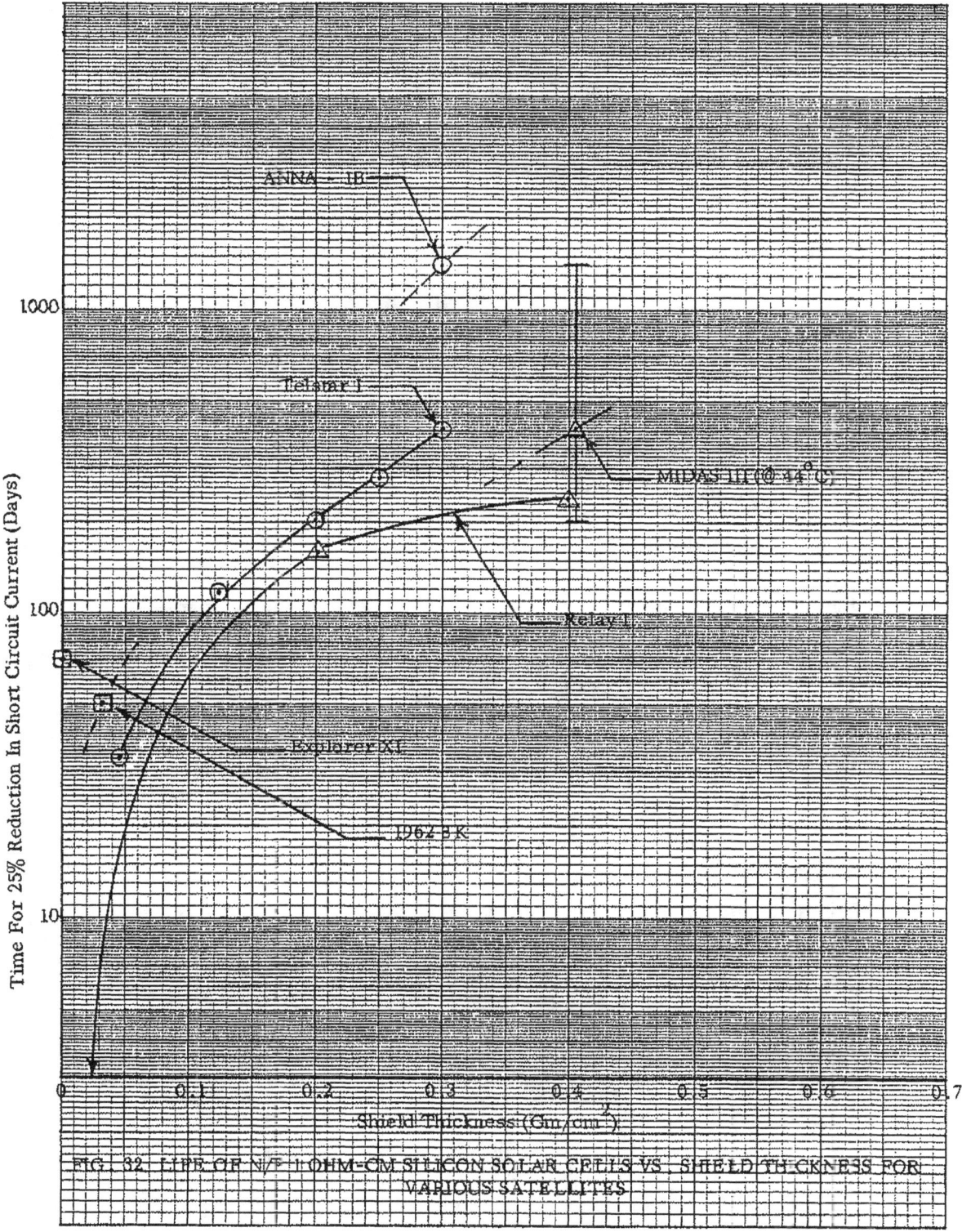

FIG. 32 LIFE OF N/P 1 OHM-CM SILICON SOLAR CELLS VS. SHIELD THICKNESS FOR VARIOUS SATELLITES

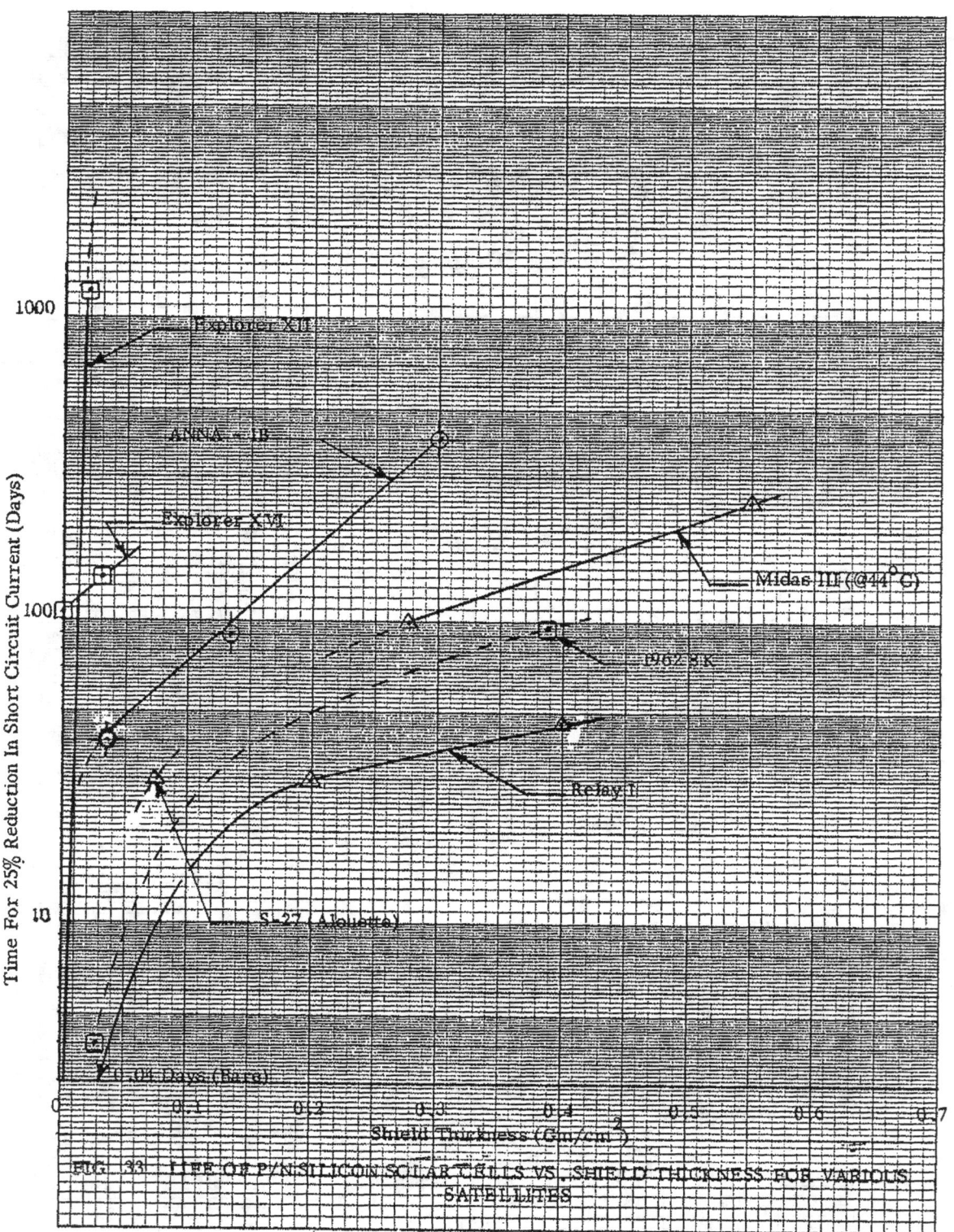

FIG. 33 LIFE OF P/N SILICON SOLAR CELLS VS. SHIELD THICKNESS FOR VARIOUS SATELLITES

0.2 gm/cm^2. Therefore, the data from Fig. 32 and 33 have been replotted in Fig. by multiplying the p/n life values by the factor 4.0. It may be seen that the shape of the resulting curve of the predicted life of n/p silicon solar cells vs. shield thickness appears to be a function of the particular satellite orbit, as expected, since it depends on the electron and proton fluxes and energy spectra encountered along the orbit. The factor of 4 difference between n/p and p/n cells is only expected to be applicable if most of the damage is produced by protons above about 2 Mev, where the damage constants are found to differ by about a factor of four. It is noted that the factor of four difference between n/p and p/n cells does not apply for the 1962 βK satellite with thin shielding where the n/p cells appear to last more than ten times as long as p/n cells. This fact may be due to one or both of two factors: (1) The p/n cells degraded rapidly in four days at a time when the flux of electrons from the Russian explosions was high, whereas the average flux during the 50 day life of the n/p cells may have been lower, and (2) even if the average flux encountered per day were constant in time, but the damage was produced mainly by electrons, a factor of the order of 20 to 100 times different life is predicted on the basis of the relative damage constants for 1 Mev electrons and a factor of 10 to 25 for 3 Mev electrons. Fig. 34 may be used as a first approximation to estimate the life of n/p 1 ohm-cm silicon solar cells in certain orbits which have orbital parameters fairly close to those for which the flight test data were obtained. It is seen that the most severe orbit with respect to radiation damage is that of Relay-1, followed in order by 1962 β K, Telstar I, Midas-III and Anna-1B. The data for the Tetrahedral Research Satellite No. 1 is not plotted because it showed no measurable degradation, even on bare p/n cells, during the first 42 days, which indicates that the damage rate to solar cells is negligible for a satellite with an apogee of only 362 nautical miles.

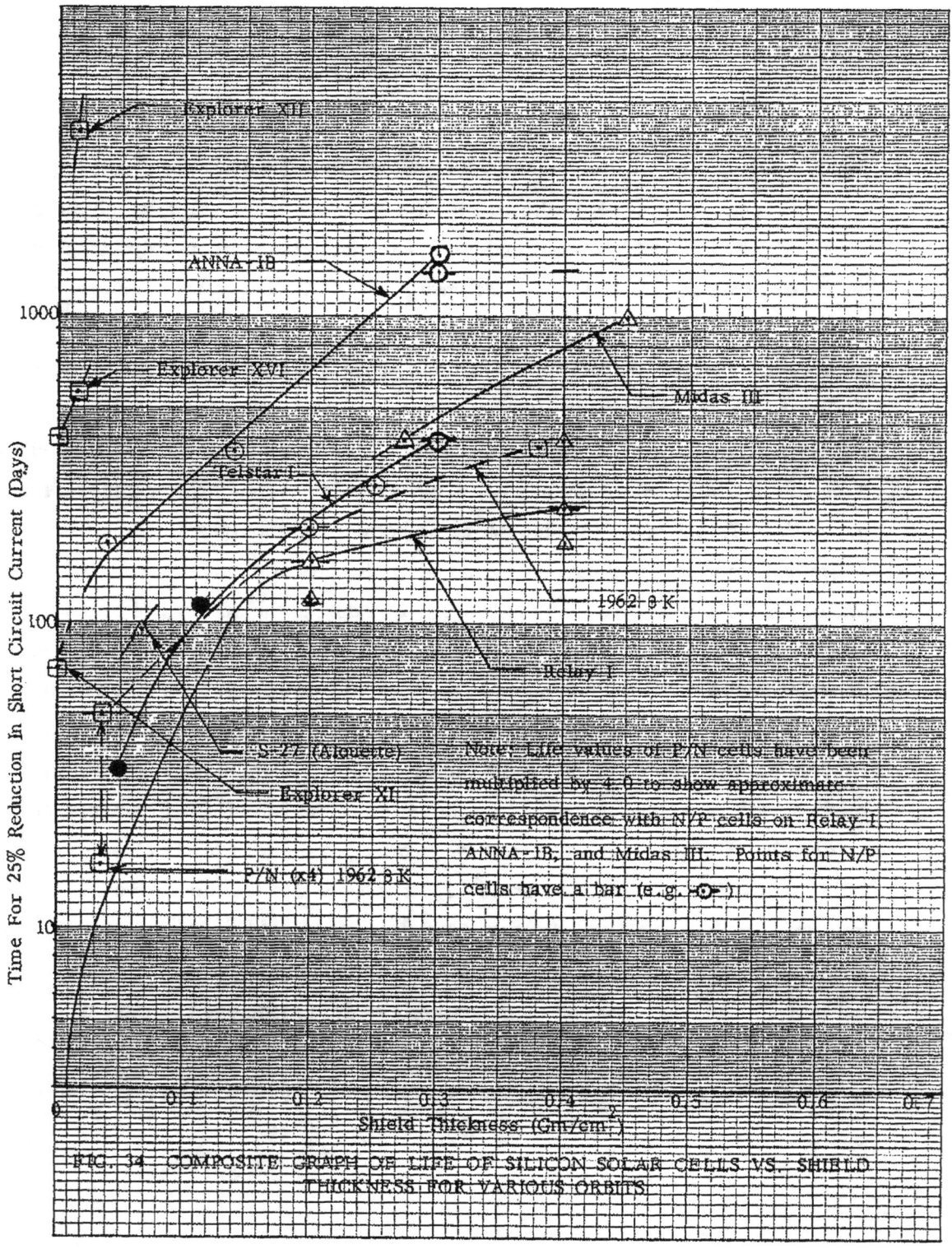

FIG. 34 COMPOSITE GRAPH OF LIFE OF SILICON SOLAR CELLS VS. SHIELD THICKNESS FOR VARIOUS ORBITS

VI DESIGN METHODS FOR SOLAR CELL POWER SYSTEMS

A. **General Requirements**

Some of the major factors which need to be considered in designing a solar cell power system for operation in a radiation environment include:

1. The severity of the radiation environment expected in the proposed orbit.
2. The availability of radiation resistant solar cells (e.g. n/p 1 ohm-cm silicon cells, higher resistivity n/p cells, or graded-base cells).
3. The thickness of shielding which can be used within the weight limit, or the increased area of cells which can be used to produce the required power within the area-limit. (Generally a tradeoff optimization study is desirable.)
4. The expected range of operating temperatures for the solar cells, which affects the selection of the solar cell operating point (the voltage per cell or the load resistance).

B. **Selection of Type of Solar Cells**

Generally if a proposed satellite is to spend an appreciable fraction of its time in the altitude zone between about 1000 km and 10,000 km, it will be necessary to use radiation resistant n/p silicon solar cells in order to attain reasonable life. In other cases there may be an economic advantage in using p/n cells unless production quantities of radiation resistant cells become available at equal cost. There is evidence that n/p silicon cells with a base resistivity greater than 1 ohm-cm and also graded-base (drift-field) cells are more radiation resistant than 1 ohm-cm cells. Therefore, it is expected that production quantities of such cells will become available for use in extreme environments. Solar cells of gallium arsenide, cadmium sulfide and other materials are still in the experimental research stages.

C. **Degradation Rates in Particular Orbits as a Function of Shield Thickness**

For certain orbits which are very similar to those of the satellites discussed in Section IV, an approximation of the time for silicon cells to degrade by 25% in short circuit current can be obtained by reference to Fig. 32, 33 and 34. For degradations greater than 25%, which are of interest in overdesigning the total cell area, it may be assumed that the short circuit current degrades by approximately 20% per decade in time. This is a good approximation only if the radiation flux encountered per day is constant, or if the time for 25% degradation is long compared to the period for apogee precession (the apogee precession period for Anna - 1B is 120 days). If one designs for 45% degradation in short circuit current, the life will be extended by a factor of 10 as compared to 25% degradation. It is clear that the use of area overdesign can be a potent method of extending solar cell life. It may be preferable in many cases to increase the number of solar cells used, rather than increase the shield thickness in order to obtain the required power at the end of life.

D. <u>Calculation of Solar Cell Degradation Rates</u>

A procedure which can be used to estimate solar cell degradation in a particular orbit is outlined below. The procedure given for calculating electron damage is only applicable for a fission electron spectrum, which is approximately correct for orbits in the region of the McIlwain L parameter less than 2 (an altitude of 4000 miles at the magnetic equator). For a more general case, changes of the electron energy spectrum along the orbit should be included in the computation.

a. Select a value of the initial diffusion length L_0 of the solar cells, typically about 120 microns (0.012 cm) for n/p one ohm-cm silicon cells.

b. Assume an allowable degradation for the short circuit current or maximum power and determine the final value of diffusion length L from Fig. 15 or 19 using the curve corresponding to short circuit current under space sunlight with the appropriate value of L_0.

c. Calculate the allowable change in $(1/L^2)$ which corresponds to the final short circuit current:

$$\Delta(1/L^2) = \left[\frac{1}{L^2} - \frac{1}{L_0^2}\right] \tag{12}$$

d. Calculation of Proton Damage Rate

1. Choose an energy distribution for the protons which is applicable to the orbit and express the flux per unit energy interval in the form:

$$\rho(E) = I_0 e^{-E/E_0}$$

This equation should apply over the range from E_1, which is the minimum proton energy capable of penetrating the assumed shield thickness as given by Fig. 35, up to approximately $E_2 = 200$ Mev.

2. Use a proton damage coefficient applicable to the solar cell base material as shown in Fig. 11 or 12.

3. Calculate the average damage integral per unit time:

$$\overline{D}_p = \frac{\Delta(1/L^2)}{t} = \frac{1}{t} \int_0^t \int_{E_1}^{E_2} K_p(E) \rho_p(E, t) \, dE \, dt \quad (11)$$

where E_1 is the minimum energy that will penetrate the shield as determined from Fig. 35. The integral with respect to time may be approximated by breaking the proton belt into several regions in B-L coordinates and calculating the time spent in each region, taking into account orbit regression and precession of the orbital apogee. In order to calculate the average damage rate, the time t should equal one period of precession of the apogee, unless it is desired to predict the degradation more precisely by integrating from the expected initial orbital parameters.

e. Calculation of Electron Damage Rate

1. Assume an energy distribution for the electrons encountered along the orbit, (e.g. a fission spectrum for orbits passing through the artificial electron belt at values of the McIlwain L parameter less than 2) and calculate the average omnidirectional flux of fission electrons per unit time (φ_F).

2. Refer to Rosenzweig's curve of equivalent 1 Mev electron flux per omnidirectional fission electron for the assumed shield thickness and solar cell type (Fig. 36) and read off the ratio R of equivalent 1 Mev integrated electron flux per unit omnidirectional fission electron flux. Calculate the equivalent average 1 Mev flux for the orbit:

$$\varphi_{E1} = R \varphi_F$$

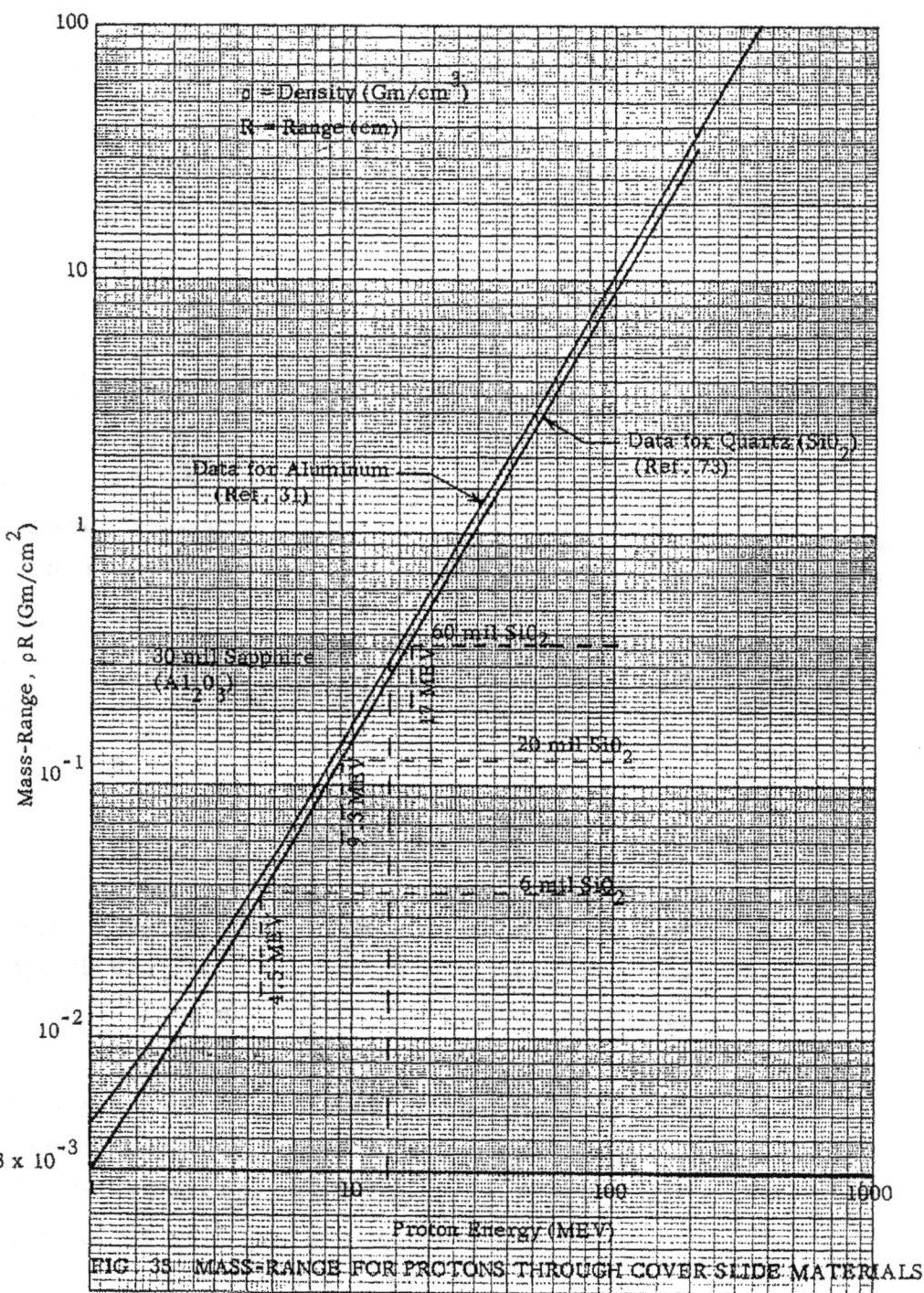

FIG. 35 MASS-RANGE FOR PROTONS THROUGH COVER SLIDE MATERIALS

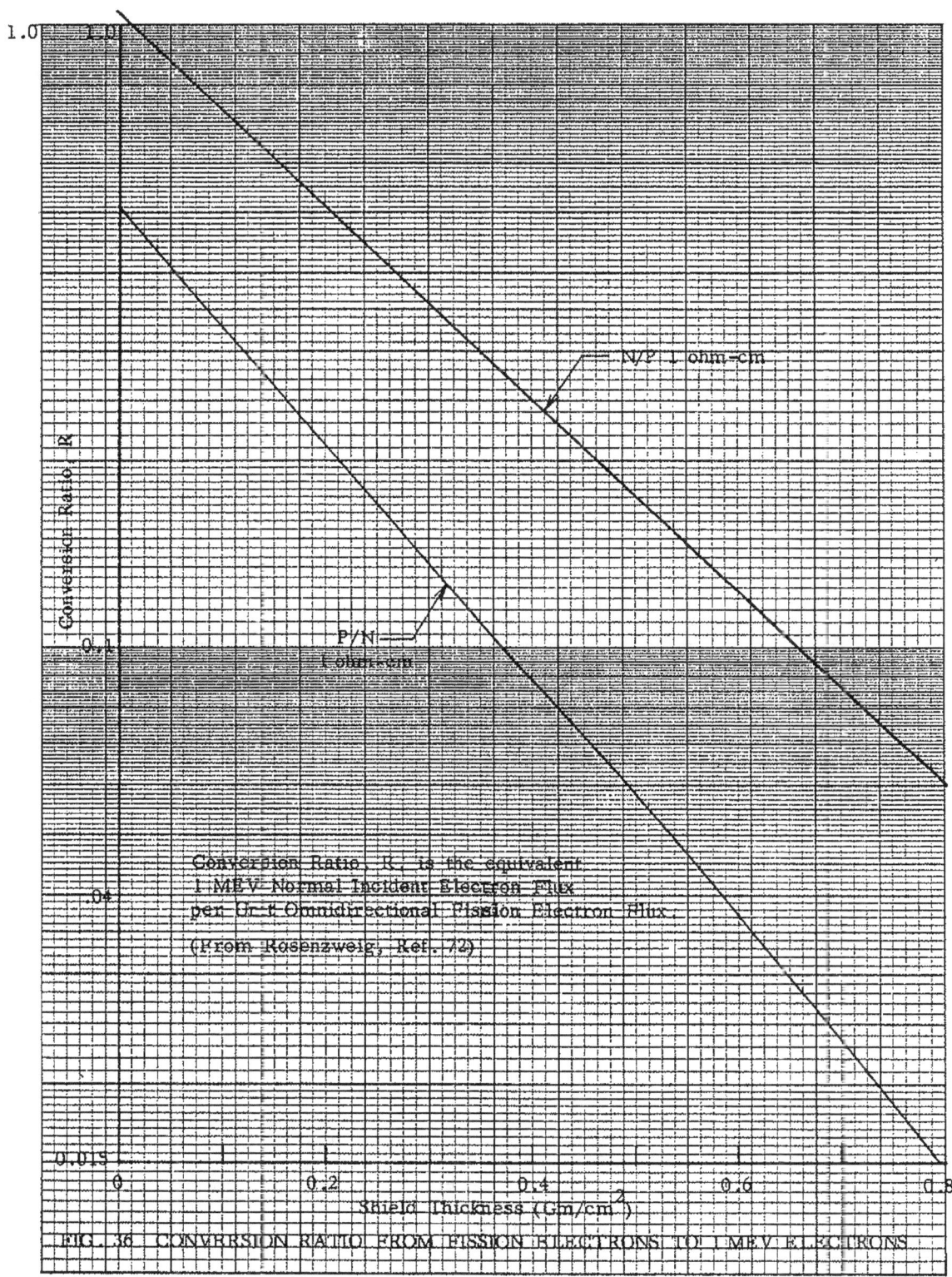

FIG. 36 CONVERSION RATIO FROM FISSION ELECTRONS TO 1 MEV ELECTRONS

f. Read the K_{E1} value for 1 Mev electrons from Fig. 8 for the type of solar cell base material used, and calculate the average electron damage rate in the orbit:

$$\overline{D}_E = K_{E1} \varphi_{E1} \qquad (13)$$

This has the units of change in $1/L^2$ per unit time.

g. Calculate the average damage rate:

$$\overline{D}_T = \overline{D}_P + \overline{D}_E \qquad (14)$$

where $\overline{D}_P$ is from Step D-3 and $\overline{D}_E$ is from Step F.

h. Solve for solar cell life to the assumed final value of short circuit current from:

$$T = \frac{\Delta (1/L^2)}{\overline{D}_T} \qquad (15)$$

where $\left[\Delta (1/L^2) \right]$ is from Step C and $\overline{D}_T$ from Step G.

E. **Sample Calculation of Radiation Damage in Equatorial Orbits**

1. Calculation of Proton Damage -

It is assumed that the proton spectral flux is given by McIlwain and Pizzella, (Ref. 100):

$$\rho(E) = I_0 e^{-E/E_0} \qquad (16)$$

where
$$E_0 = 306 \, L^{-5.2} \, (\text{Mev}) \qquad (17)$$
$$L = \text{McIlwain's parameter}$$

For a satellite at an altitude of 2000 miles (half an earth radius) $L = 1.5$ and $E_0 = 37.3$ Mev. The value of I_0 is found by normalizing to an integral flux above 40 Mev. of 2×10^4 protons/cm^2-sec, as given by O'Brien (Ref. 104):

$$I_0 \int_{40}^{\infty} e^{-E/E_0} \, dE = 2 \times 10^4 \qquad (18)$$

or $\quad I_0 = 1570 \text{ p/cm}^2$ - sec - Mev

The radiation damage to n/p 1 ohm-cm silicon solar cells covered by 30 mils of sapphire shielding on the front and with infinite back shielding will be estimated. The proton damage coefficient is assumed to be

$$K_p = A/E$$

where
$$A = (1.5 \times 10^{-5}) \text{ for } E < 60 \text{ Mev} \quad (19)$$
$$K_p = B = 2.5 \times 10^{-7} \text{ for } E > 60 \text{ Mev}$$

It is assumed that the energy spectrum and flux of protons behind the shield is the same as on the outside of the shield except that the low energy particles are stopped by the shield. This assumption is equivalent to assuming that the lower energy and therefore higher damage coefficient for those protons which penetrate the shield is just compensated by the fact that an omnidirectional flux of incident protons is strongly absorbed at small angles of incidence so that only protons within a certain entrance cone can penetrate. This assumption is conservative in that it somewhat over estimates the damage produced by protons which have energies slightly greater than that required to penetrate the shield. (Compare Rosenzweig's calculated results as shown in Fig. 37 The change in inverse squared diffusion length for the cells is calculated from:

$$\Delta(1/L^2) = \frac{T}{2} \int_{E_c}^{\infty} \rho(E) \, K(E) \, dE \quad (20)$$

where T = time of exposure, seconds

E_c = cutoff energy = 14 Mev for 30 mil sapphire (0.3 gm/cm^2)

or

$$\Delta(1/L^2) = \frac{I_0 T}{2} \left[\int_{E_c}^{60} A E^{-1} e^{-E/E_0} \, dE + \int_{60}^{\infty} B e^{-E/E_0} \, dE \right] \quad (21)$$

If we let $x = \frac{E}{E_0}$

Then

$$\Delta(1/L^2) = \frac{I_0 T}{2} \left[A \int_{x_c}^{x_{60}} \frac{1}{x} e^{-x} \, dx + B E_0 \int_{x_{60}}^{\infty} e^{-x} \, dx \right] \quad (22)$$

where $x_c = \frac{14}{E_0}$ $x_{60} = \frac{60}{E_0}$

and

$$\int_{x_{60}}^{\infty} e^{-x} \, dx = e^{-60/E_0} \quad (23)$$

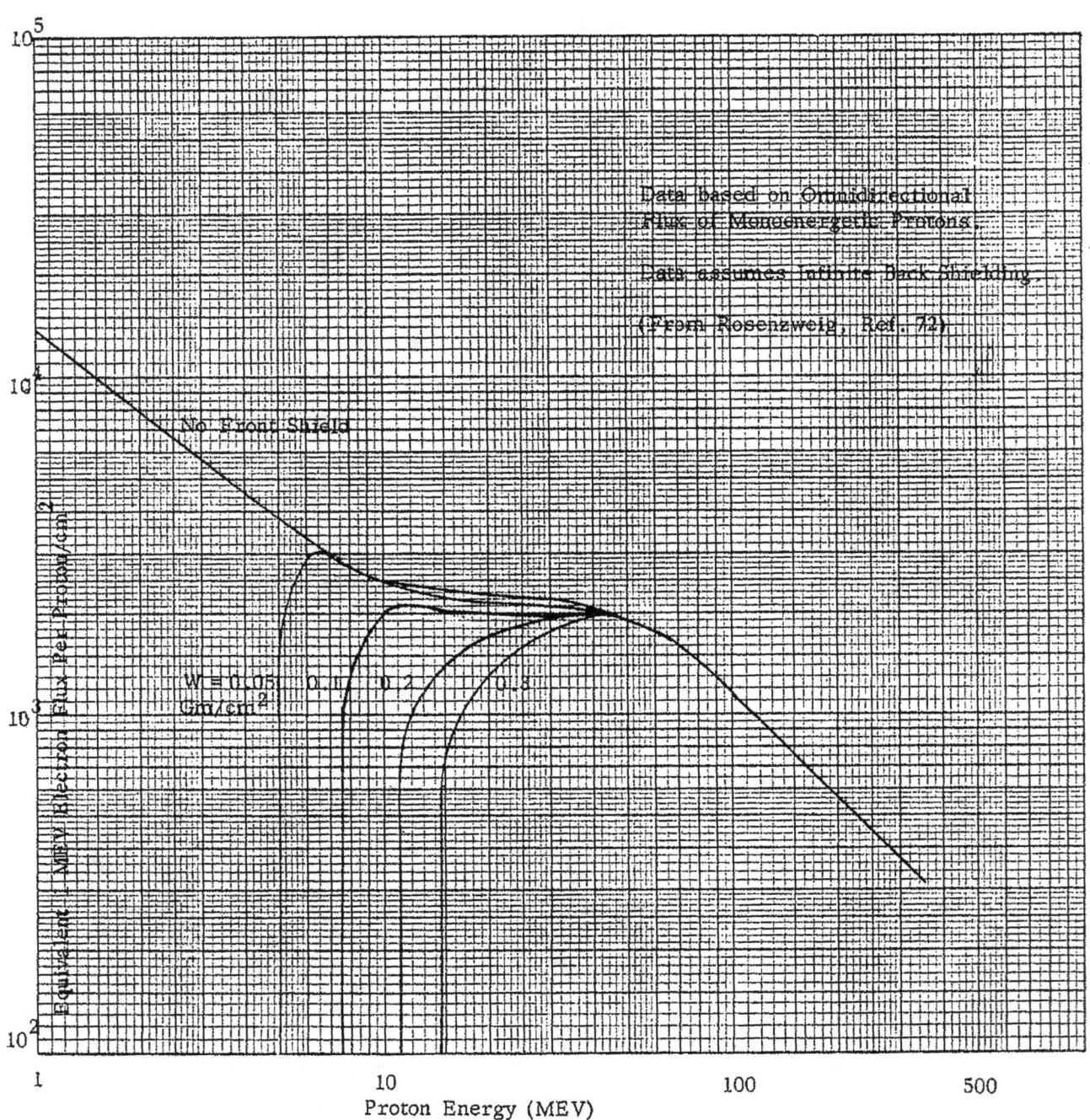

FIG. 37 DAMAGE RATE BY PROTON FLUX AS A FUNCTION OF PROTON ENERGY

The first integral in equation 22 may be evaluated as an infinite series:

$$F(x) = \int \frac{e^{-x}}{x} dx = \ln x - x + \frac{x^2}{2 \cdot 2!} - \frac{x^3}{3 \cdot 3!} + \frac{x^4}{4 \cdot 4!} \quad (24)$$

A plot of the function $F(x)$ is shown in Fig. 38. Using this figure we evaluate (with $E_0 = 37.3$):

$$F(x_{60}) - F(x_{14}) = F(1.61) - F(0.375) = -0.62 - (-1.30) = 0.68 \quad (25)$$

Evaluating the damage rate, we have:

$$\frac{\Delta(1/L^2)}{T} = \frac{I_0}{2} \left[1.02 \times 10^{-5} + 0.186 \times 10^{-5} \right] = 6 \times 10^{-6} I_0 \quad (26)$$

It is seen that the damage by protons above 60 Mev is only 18% of that by protons from 14 to 60 Mev for the assumed spectral distribution.

For $I_0 = 1570$ p/cm^2-sec-Mev

$$\frac{\Delta(1/L^2)}{T} = 9.4 \times 10^{-3} \text{ cm}^{-2} \text{ sec}^{-1}$$

For solar cells with an initial diffusion length of 120μ, a decrease of 25% short circuit current is obtained when L is about 15.5μ. The time required to degrade to L = 15.5μ by protons only is:

$$T_p = \frac{\frac{1}{(15.5)^2} - \frac{1}{(120)^2}}{9.4 \times 10^{-3}} 10^8$$

$$T_p = \frac{4.10 \times 10^5}{9.4 \times 10^{-3}} = 4.35 \times 10^7 \text{ sec} = 505 \text{ days}$$

This is the calculated life under proton damage to reach 75% of initial current for n/p 1 ohm-cm cells with 30 mils of sapphire shielding in a circular equatorial orbit at 2000 miles altitude.

2. Calculation of Electron Damage -

For equatorial orbits between L = 1.3 and 1.8 (1200 to 3200 miles altitude) it is assumed that a fission electron spectrum applies with an omnidirectional flux of 5×10^8 e/cm^2 sec, which is one half the value given by Hess for a time of one week after Starfish. (This is probably too high a flux and too hard a spectrum at L=1.8, based on personal communication from Walter L. Brown.)

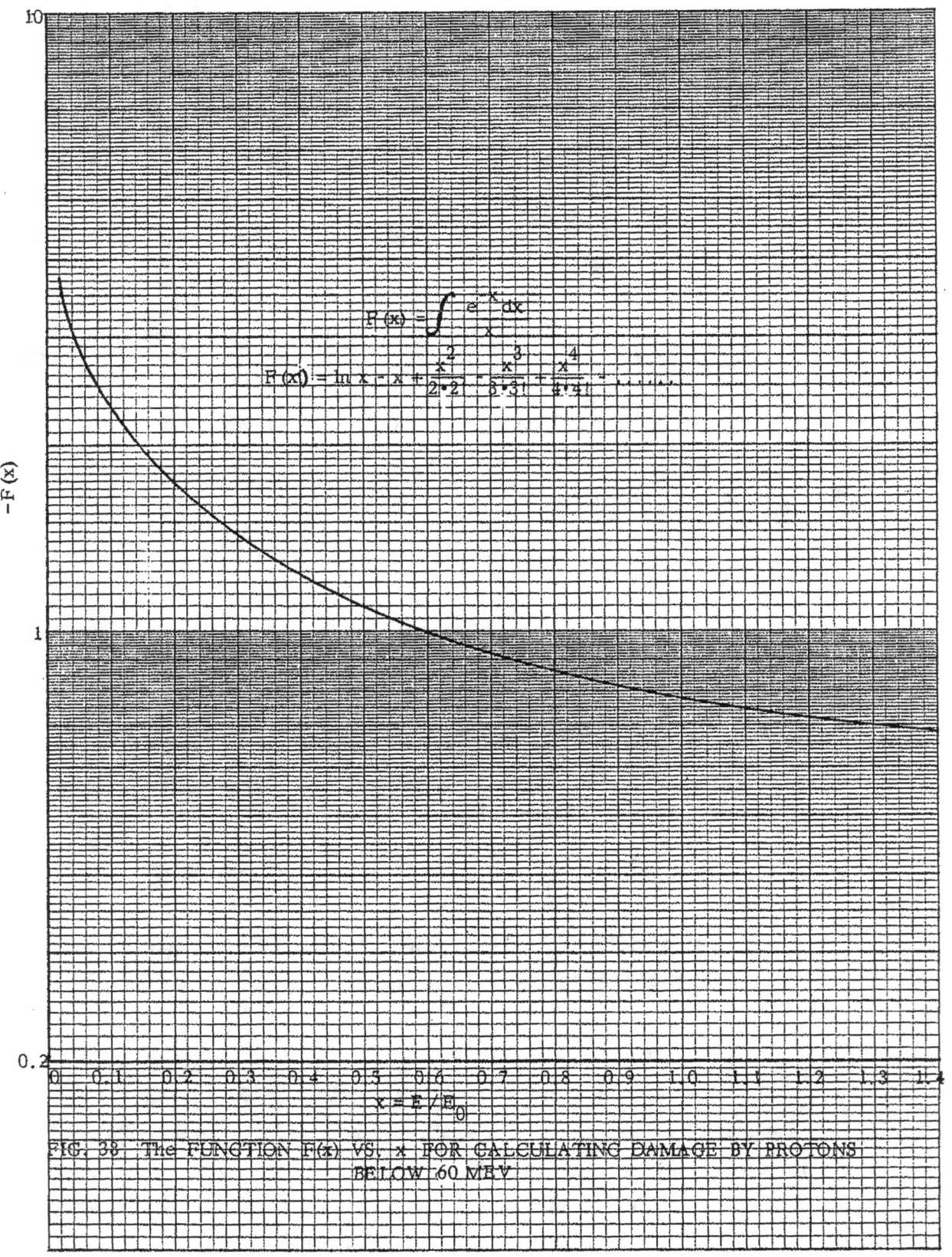

FIG. 38 The FUNCTION F(x) VS. x FOR CALCULATING DAMAGE BY PROTONS BELOW 60 MEV

From Rosenzweig's data (Fig. 36) the conversion factor from omnidirectional fission electrons to equivalent normal incident 1 Mev electrons is 0.37 for a shield thickness of 0.3 gm/cm^2 on n/p cells.

The change in $1/L^2$ per 1 Mev electron is found to be approximately

$$K_{E1} = \frac{4.1 \times 10^5}{2.5 \times 10^{15}} = 1.64 \times 10^{-10}$$

(Fig. 8 shows $K_{E1} = 1.7 \times 10^{-10}$ at 1 Mev)

The average electron damage rate is:

$$\overline{D}_E = K_{E1}\, \varphi_{E1} = (1.64 \times 10^{-10})(0.37)(5 \times 10^8)$$

$$\overline{D}_E = 0.0303 \text{ cm}^{-2} \text{ sec}^{-1}$$

The time to degrade the cells by 25%, considering only electrons is:

$$T_E = \frac{\Delta(1/L^2)}{\overline{D}_T} = \frac{4.1 \times 10^5}{0.0303} = 1.35 \times 10^7 \text{ sec} = 156 \text{ days}$$

3. Calculation of Combined Effect of Electrons and Protons -

The total damage rate is

$$D_T = D_E + D_P = \frac{\Delta(1/L^2)}{T_T} = \frac{\Delta(1/L^2)}{T_E} + \frac{\Delta(1/L^2)}{T_P} \quad (27)$$

from which

$$\frac{1}{T_T} = \frac{1}{T_E} + \frac{1}{T_P} \quad (28)$$

or

$$T_T = \frac{T_E T_P}{T_E + T_P} \quad (29)$$

The life under combined electrons and protons is then

$$T_T = \frac{(156)(505)}{(156 + 505)} = 119 \text{ days}$$

This is the calculated life to reach 75% of initial current for n/p 1 ohm-cm silicon cells with 30 mils of sapphire shielding in a circular equatorial orbit at 2000 miles altitude.

Similar calculations have been made for other altitudes using the proton flux map shown in Fig. 30 and the results are plotted in Fig. 39. It is seen that the damage rate by protons increases with altitude and has not reached its maximum at an altitude of 5000 km, above which the absolute value of proton fluxes are not well defined. This effect is due to the increasing fluxes of protons with energies slightly above the 14 Mev shield cutoff energy which are encountered with increasing values of the McIlwain L parameter.

The solar cell life of 80 days calculated for an altitude of 5100 km may be compared with the observed life of 190 days interpolated from Relay-1 flight test data. One may conclude that the Relay-1 damage rate is equivalent to spending about 40% of its time at the most damaging point calculated, namely at 5100 km altitude at the equator. Actually Relay spends considerable time at larger L values than 1.8 where it encounters many low energy protons and it also passes through the artificial electron belt. Therefore, a computer calculation would be required after the fluxes and spectra are better defined in order to compare predicted with observed degradation rates.

F. Comparison of Calculated and Observed Damage on S-27 Alouette.

The Alouette is in approximately a 1000 km circular polar orbit for which Hess (Ref. 39) has calculated an omnidirectional flux of fission electrons of 2×10^{12} per day at a time of one week after Starfish. Alouette was actually launched on Sept. 29, 1962, nearly three months after Starfish. For blue-shifted p/n solar cells with 12 mils (0.076 gm/cm^2) of glass shielding, Rosenzweig's conversion ratio from Fig. 36 is found to be 0.37, so the equivalent 1 Mev flux is 0.74×10^{12} e/cm^2 day.

The damage coefficient for p/n cells at 1 Mev is selected from Fig. 9 to be 1.22×10^{-8} and the change in $1/L^2$ to produce 25% reduction in short circuit current is taken as 4.1×10^5 cm^{-2}. Therefore, the time to reach 25% reduction in short circuit current under electron damage is

$$T_E = \frac{\Delta(1/L^2)}{K\varphi}$$

$$T_E = \frac{4.1 \times 10^5}{(1.22 \times 10^{-8})(0.74 \times 10^{12})} = 45 \text{ days}$$

This is probably an underestimate of "electron life" because of decay of the artificial belt.

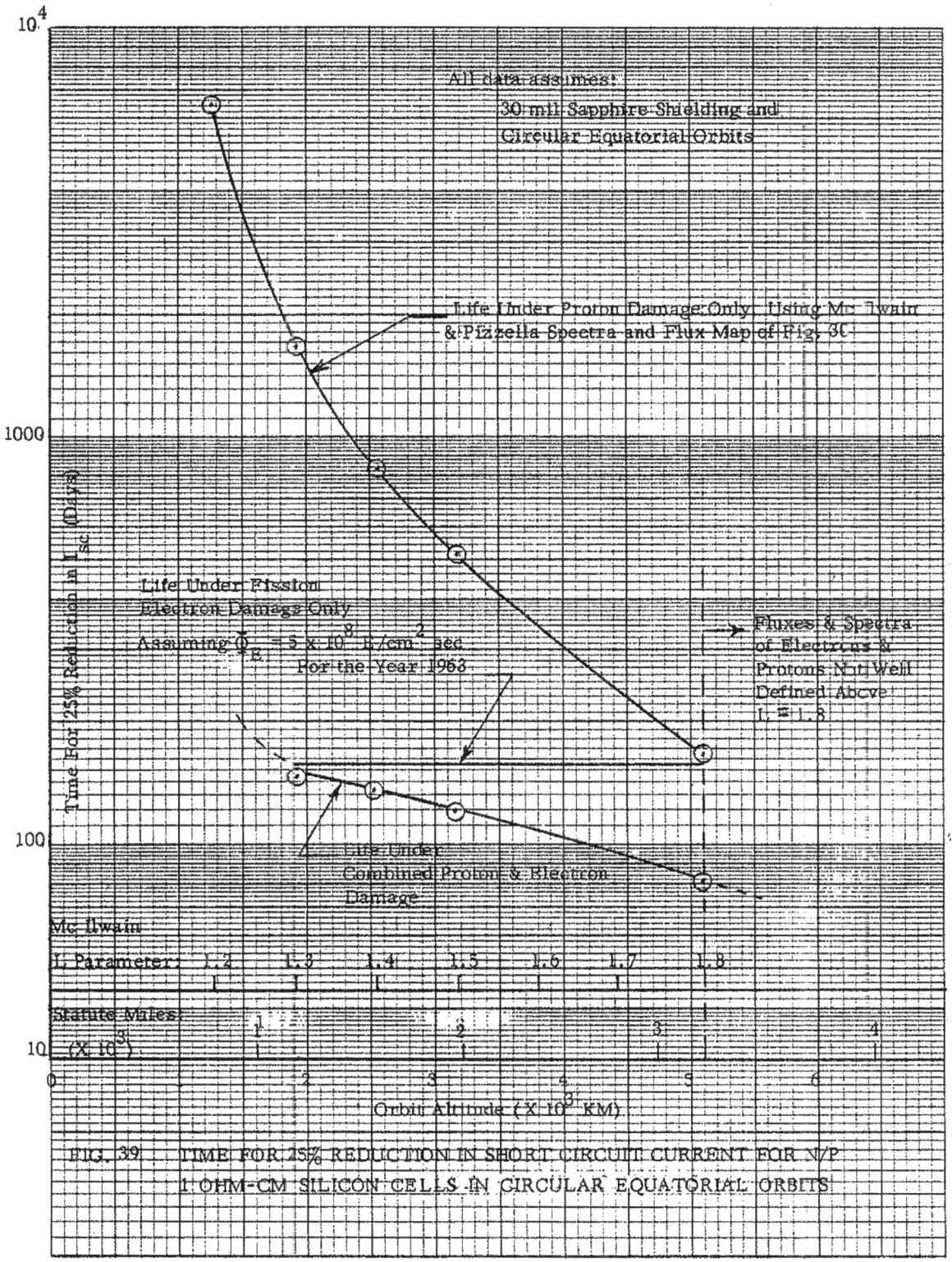

FIG. 39 TIME FOR 25% REDUCTION IN SHORT CIRCUIT CURRENT FOR N/P 1 OHM-CM SILICON CELLS IN CIRCULAR EQUATORIAL ORBITS

Flight test data from Alouette showed a 25% degradation after about 30 days. The added damage sustained may be accounted for by protons encountered in the South Atlantic Anomaly.

APPENDIX A
RADIATION EFFECTS ON SOLAR CELL COVER SLIDE MATERIALS AND ADHESIVES

A. <u>Introduction</u>

The prediction of solar cell power system life should take into account the degradation of the optical transmissive properties of the cover materials. The covering materials are generally a transparent cover slide and a transparent adhesive. The cover slide is utilized to shield against electrons and protons, and to carry a spectrally selective filter. It also protects against contamination from being handled on the ground. Investigations have been conducted with protons and electrons on various cover slide materials and adhesives. From these studies an indication of the degradation that can be expected in service may be derived.

B. <u>Damage Mechanism</u>

Cover materials are susceptible to "F" center formation from high energy particle fluxes. The formation of these color absorbing centers is a function of the impurities in the material. Most color centers can be annealed out of the material by heat, infrared radiation or visible light.

In general, the sensitivity to coloration is not fixed for a given type of engineering material but varies according to the impurities, lattice imperfections, strain, stoichiometry and the prior deformation of the material (Ref. 118). When irradiation is at elevated temperatures an equilibrium density of color centers is attained from the production and destruction of the color centers.

Relatively little work has been done on coloration or the coloration equilibrium from simultaneous energetic protons, electrons and ultraviolet radiation below 3000 Angstroms in the presence of the longer wavelength components of sunlight. However, the change in transmittance of various materials from proton and electron radiation has been measured by F. A. Campbell of the Naval Research Laboratory (Ref. 112) and by J. L. Patterson and G. A. Haynes (Ref. 120) of the Langley Research Center. These data give an indication of the amount of damage that can be expected in service.

C. Cover Slide Requirements

The most desirable materials for solar cell cover slides are those which show the least amount of transmission degradation within the solar cell spectral response band upon exposure to electron and proton radiation. This spectral band is typically 0.35 to 1.1 microns, with a peak near 0.8 microns. When the spectral distribution of space sunlight is taken into consideration, the peak energy point occurs at approximately 0.6 microns.

D. Electron and Proton Damage Experiments

Assessment of electron damage has been made by F. A. Campbell (Ref. 112 and 113) of the Naval Research Laboratory. Electrons of 1 Mev energy at a total dose of 10^{16} e/cm^2 were utilized. Spectral transmittance was measured with a Beckman IR-4U double beam spectrophotometer. The electron exposure was conducted in air, directly under the beam of a Van de Graaff. Transmittance values for various materials between 0.4 and 0.7 microns are summarized in Table A-1 (Ref. 113).

Patterson and Haynes (Ref. 120) utilized a Radiation Dynamics No. P.E.A.-10 accelerator with a constant beam current density of 0.03 microamp/cm^2. Electron energy of 1.2 Mev was used and most of the samples were exposed to a total dose of 2.7×10^{15} e/cm^2. A 2900°K filament temperature tungsten lamp was used as the source and the transmission efficiency was recorded as a percent change in broad spectral band light passing through the sample to a typical solar cell. The measurement accuracy was 1/2%.

Measurements of changes in spectral transmission after proton irradiation were conducted at the Naval Research Laboratory. The irradiation consisted of a single dose of 4×10^{11} p/cm^2 at an energy of 4.6 Mev. The samples were irradiated in vacuum at a pressure of 10^{-4} mm Hg and the results are summarized in Table A-2. The test data show a superiority for the Corning 7940 specimens as compared to Corning 0211.

1. Silica

Two major classes of silica are synthetic quartz (or fused silica) and vapor deposited silica. Both electron and proton damage is reported for various types and manufacturers of fused silica as shown in Tables A-1, A-2 and A-3. Electron radiation

TABLE A-1

Effects of 1 Mev Electrons On Spectral Transmittance (T) Of Solar Cell Cover Materials (Ref. 112 and 113)

1 Mev Electrons - Total Dose of 10^{16} e/cm^2

Material Description		50% cut-off point µ	Wavelength in Microns				50% cut-off point µ	NOTES
			0.40	0.50	0.60	0.70		
1. Microsheet, Corning 0211, 6 mil	T_0		89.0	90.0	90.0	90.0		
	T_F		82.0	85.0	87.0	99.0		
	%ΔT		7.9	5.6	3.3	2.2		
2. Same as (1) + antireflecting coating + "Blue" filter	T_0	0.444	Below 50% cut-off	96.0	97.5	96.0		
	T_F	0.444	"	89.0	92.5	92.0		
	%ΔT			7.3	5.1	4.2		
3. Same as (1) + antireflecting coating + "Blue-red" filter	T_0	0.444	Below 50% cut-off	91.0	93.0	94.0	1.136	*
	T_F	0.442	"	82.0	85.0	86.0	1.138	
	%ΔT			9.9	8.6	6.9		
4. Microsheet, Corning 0211, 3 mil + antireflecting coating + "Blue" filter	T_0	0.435	Below 50% cut-off	93.0	94.0	-		Parallels #2 with a change only in %ΔT
	T_F	0.442	"	89.5	92.0	-		
	%ΔT			3.7	2.1			
5. Fused Silica, Corning 7940, 66 mil	T_0		89.0	90.0	90.0	90.0		*
	T_F		88.5	88.5	88.0	89.0		
	%ΔT		0.6	1.7	2.2	1.1		

TABLE A-1 cont'd

Material Description		50% cut-off point μ	Wavelength in Microns				50% cut-off point μ	NOTES
			0.40	0.50	0.60	0.70		
6. Fused Silica, Corning 7940, 20 mil + antireflecting coating + "Blue" filter	T_0	0.416	Below 50% cut-off	89.0	90.0	91.0		*
	T_F	0.425	"	88.0	88.0	90.0		
	% ΔT			1.1	2.2	1.1		
7. Non-browning lime glass Corning 8365, density 2.7, 60 mil	T_0	0.365	0.45 μ 89.0	89.0	89.0	89.0		radiation caused internal crazing in specimen, diffusing light
	T_F	-	29.0	30	33	35		
	% ΔT							
8. Non-browning lead glass Corning 8365, density 3.3, 60 mil	T_0	0.380	72.0	73.0	74.0	74.0		
	T_F	--	21.0	22.0	22.0	24.0		
	% ΔT							
9. High density lead glass	T_0	0.392	82.0	84.0	85.0	84.0		
	T_F	0.402	72.0	78.0	82.0	83.0		
	% ΔT		12.2 / 0.40 μ	7.1	3.5	1.2		
10. Adhesives ES-10, Spectrolab	T_0		89.0	90.0	90.0	90.0		Cast 1-2 mil thin with Corning 7940 6 mil fused silica cover. Crazing and delaminating visible.
	T_F		88.5	88.5	88.5	89.0		
	% ΔT		0.6	1.7	1.7	1.1		
11. Adhesive 15-3, Furane	T_0		89.0	90.5	90.0	91.0		Cast 1-2 mil thin with 6 mil Corning 7940 fused silica cover
	T_F		38.0	69.0	78.0	80.0		
	% ΔT		57.4	24	13	12		

TABLE A-1 cont'd

Material Description		50% cut-off point μ	Wavelength in Microns				50% cut-off point μ	NOTES
			0.40	0.50	0.60	0.70		
12. Adhesive DER-332 (LC) (Dow)	T_0		63	65	66	67		Cast 1-2 mil thin layer covered with 66 mil thick sheet of fused silicia
	T_F		37	53	60	64		
	%ΔT		41.3	8.6	9.1	4.5		

* Transmittance reading taken at nearest fringe peak to the wavelength indicated on all filter covered slides

T_0 - Initial preirradiated transmittance

T_F - Transmittance after 10^{16} e/cm^2

% ΔT - Percentage decrease in transmittance

TABLE A-2

Effects Of 4.6 MEV Protons On Spectral Transmittance (T) Of Solar Cell Cover Materials (Ref. 112 & 113)

Total Dose = 4×10^{11} p/cm^2

Material Description		50% cut-off point μ	Wavelength In Microns					NOTES
			0.40	0.45	0.50	0.60	0.70	
1. Microsheet, Corning 0211, 6 mil	T_O		89.5	89.5	89.5	89.5	90	
	T_F		83.5	85.0	86.5	88.5	89.0	
	% ΔT		6.7	5.0	3.4	1.1	1.1	
2. Sames as (1) + Antireflecting coating + "blue" filter	T_O	0.443	Below 50% cut-off	93.5	96.5	97.0	-	
	T_F	0.446		88.0	91.5	95.0	-	
	% ΔT			5.9	5.2	2.1	1.1	
3. Fused silica Corning 7940 30 mil	T_O		79.0	80.0	no change	no change	no change	
	T_F		80.0	80.0	"	"	"	
	% ΔT		-1.2	no change	"	"	"	
4. Fused Silica Corning 7940 20 mil + antireflecting coating + "blue" filter	T_O	0.425						
	T_F	0.427						
	% ΔT		"	"	"	"	"	

TABLE A-3

Effects of 1.2 Mev Electron Radiation on Transparent Materials (From Patterson & Haynes, Ref. 120)

Manufacturers	Type number or trade name	Sample thickness, in.	Total integrated flux, electrons/cm^2	Percent decrease in light transmission
	Fused Silica			
Corning Glass Works	No. 7940 (UV grade)	1/16, 1/8	2.7×10^{15}	0.0
Corning Glass Works	No. 7940 (optical grade)	1/8	2.7×10^{15}	0.0
Engelhard Industries, Inc. Amersil Quartz Div.	Suprasil II	1/16, 3/32	"	0.0
"	Optical	1/16	"	1.8
"	Homosil	1/16	"	2.1
"	Ultrasil	1/16	"	6.4
"	Infrasil II	1/16	"	23.0
Thermal American Fused Quartz Co.	Spectrasil A	1/8	"	0.0
General Electric Co. Willoughby Quartz Plant	GE 104	3/32	"	0.8
"	GE 105	3/32	"	30.0
"	GE 106	3/32	"	26.6
Dynasil Corporation	Dynasil Optical Grade	1/8	"	0.0
	Other Materials			
Linde Company	Sapphire	0.020	"	0.0
Pittsburgh Plate Glass Co.	Solex	1/4	"	2.7
Corning Glass Works	Vycor	1/4	1.7×10^{15}	58.9
--	Soda Lime Plate Glass	1/4	"	26.0

damage curves for Corning 0211 (microsheet) with various types of filters and antireflection coatings are plotted separately in Fig. A-1, A-2, and A-3 between wavelengths of 0.35 to 1.20 microns. The spectral degradation of transmittance for electron and proton irradiation of Corning 0211 is predominately at the shorter wavelengths and decreases gradually with increasing wavelength. The "blue-red" filter specimen however is an exception and in Figure A-3 shows appreciable degradation at all wavelengths up to about 1.0 micron. The microsheet (Corning 0211) seems to approach a saturation limit since it was found that the transmittance at all wavelengths was only slightly lower at 10^{16} e/cm^2 than at the 10^{14} e/cm^2 level.

Electron radiation damage for Corning 7940 fused silica between 0.40 and 0.70 microns is summarized in Table A-1, numbers 5 and 6. The 7940 specimen showed 2.2% or less change in transmittance for both types of particle irradiation and was the best type of cover material reported.

Table A-3 summarizes the decrease in transmittance of fused silica with 1.2 Mev electrons at a total dose of 2.7×10^{15} e/cm^2. Corning 7940, Suprasil II, Spectrasil A, and Dynasil Optical Grade had no detectable degradation and in each case, were made by a vapor deposition process. An example of the scatter in transmittance values as a function of purity of silica as shown in Fig. A-4. The poorest silica had 50 to 70 parts per million of impurities and showed close to a 40% decrease in transmission. The vapor-deposited silica had less than 1 part per million of impurity and showed a superiority over the fused crystalline silica. Those silica samples that darkened in Table A-3 were made by the process of fusing crystalline quartz.

In general, in each case in Table A-3 where darkening was noted, the curves indicated a broad absorption band centered at approximately 5500 $A^°$. Therefore, if space sunlight illumination had been used with a solar cell, the decrease of useful light transmission would have been greater than indicated.

Commercial-grade fused silica, 1/16 inch thick, from the Amersil Division of Engelhard Industries, Inc. was irradiated with protons in conjunction with a solar cell experiment at Langley Research Center (Ref. 116). There was no darkening of the silica with a 5×10^{11} p/cm^2 dose of 240 mev protons, but with a 2×10^{13} p/cm^2 dose of 22 Mev protons the cover slide darkening reduced the short circuit current by 3.7% for one sample

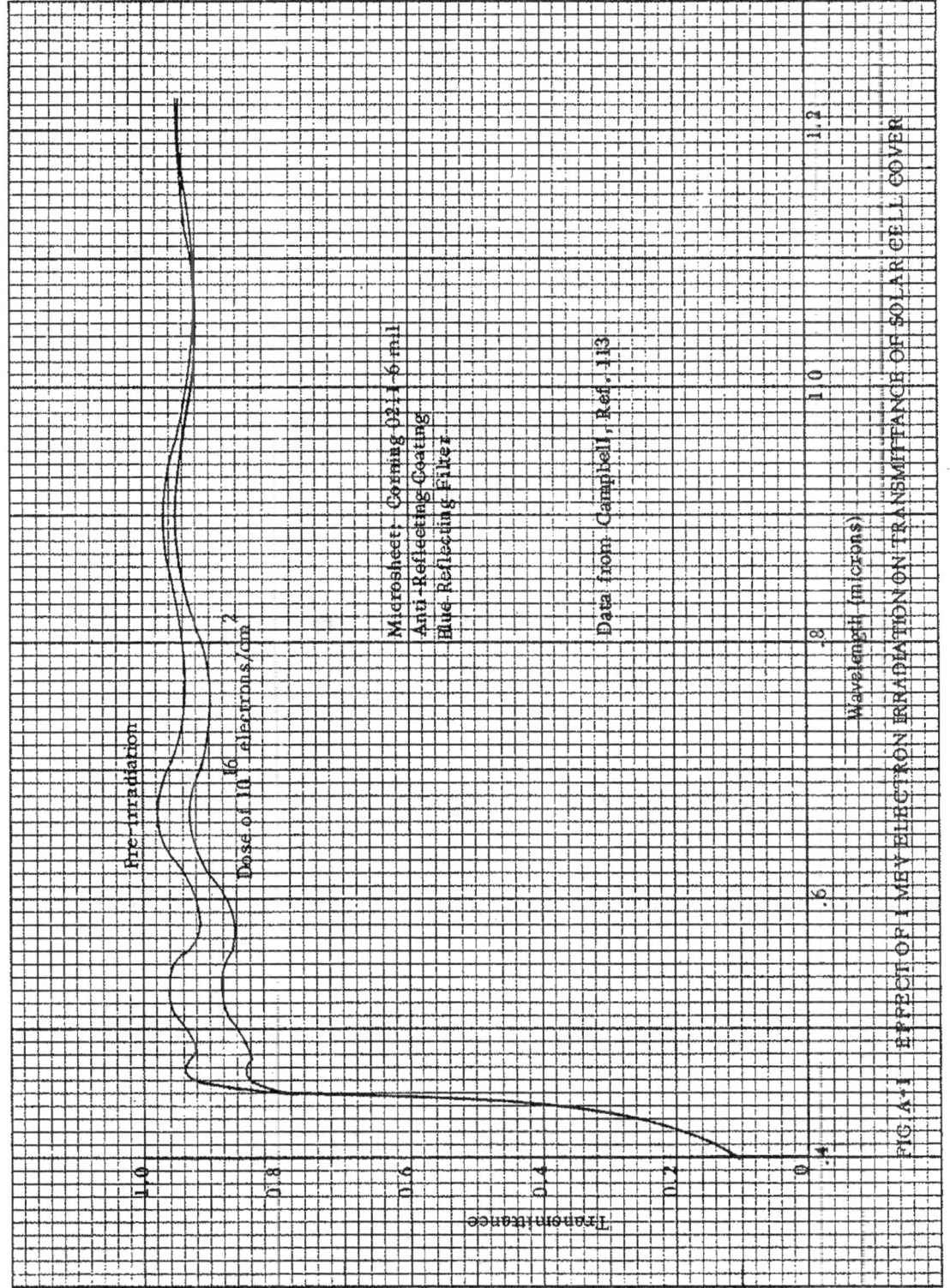

FIG. A-1 EFFECT OF 1 MEV ELECTRON IRRADIATION ON TRANSMITTANCE OF SOLAR CELL COVER

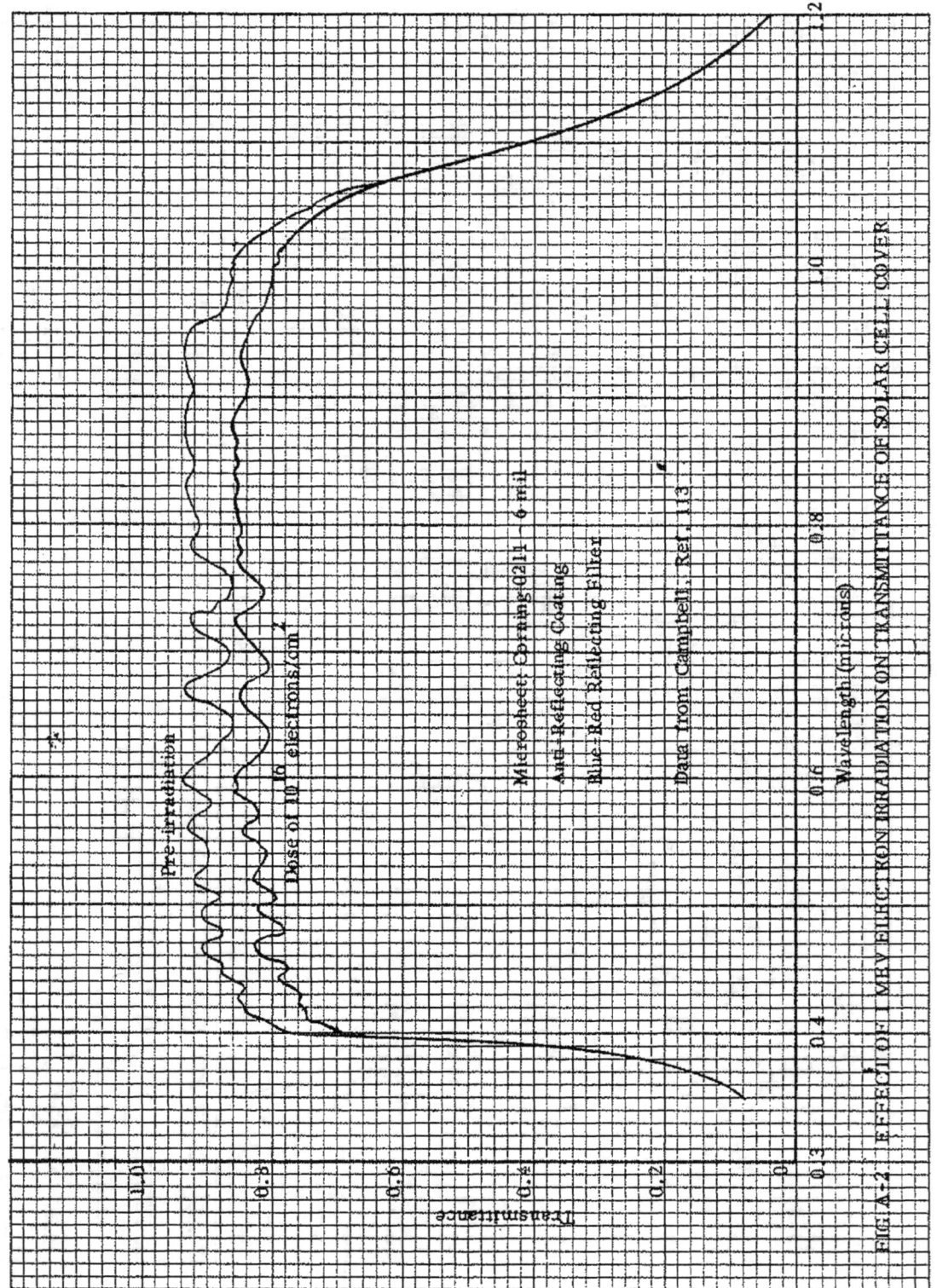

FIG. A-2: EFFECT OF 1 MEV ELECTRON IRRADIATION ON TRANSMITTANCE OF SOLAR CELL COVER

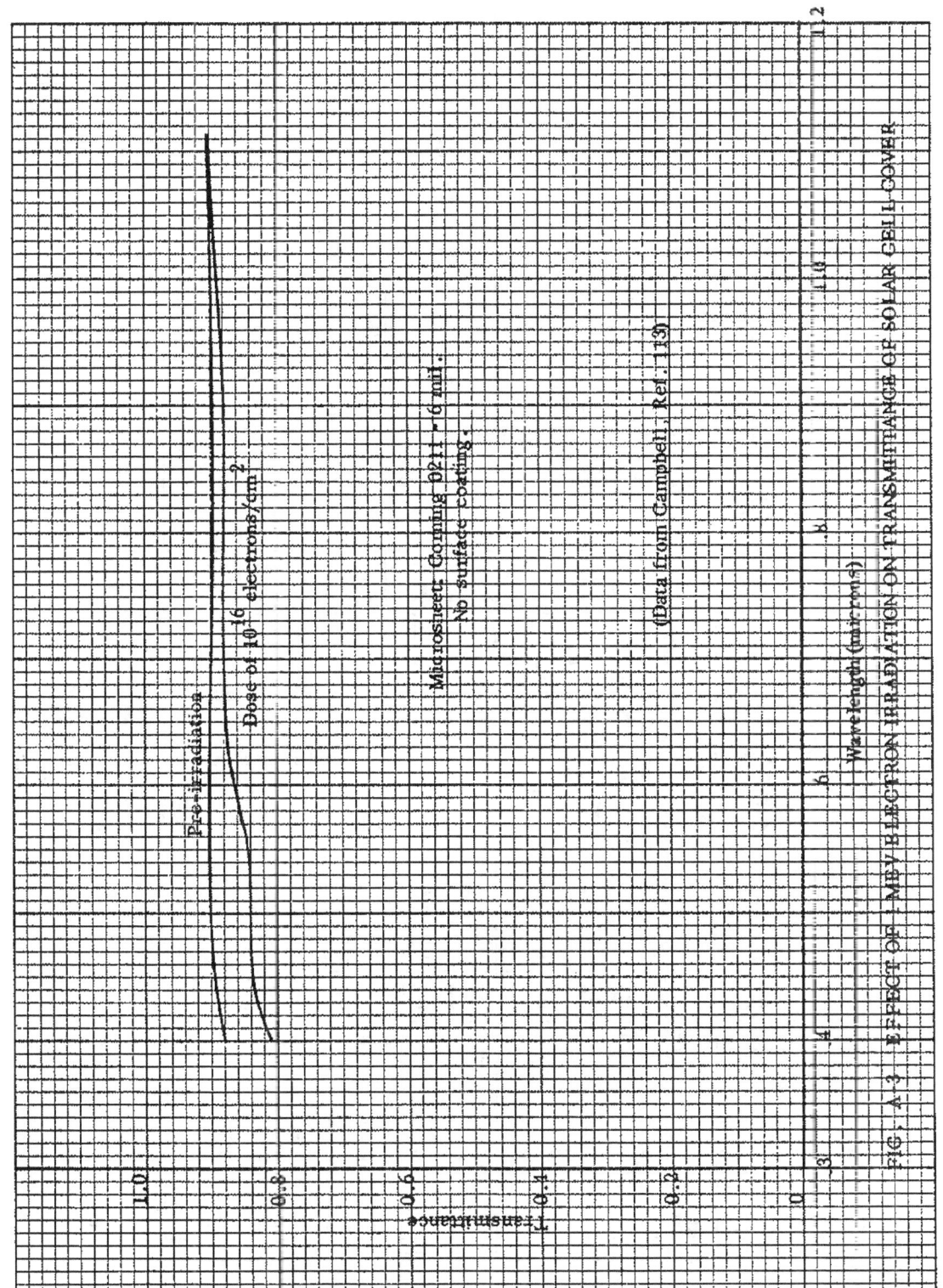

FIG. A-3 EFFECT OF 1 MEV ELECTRON IRRADIATION ON TRANSMITTANCE OF SOLAR CELL COVER

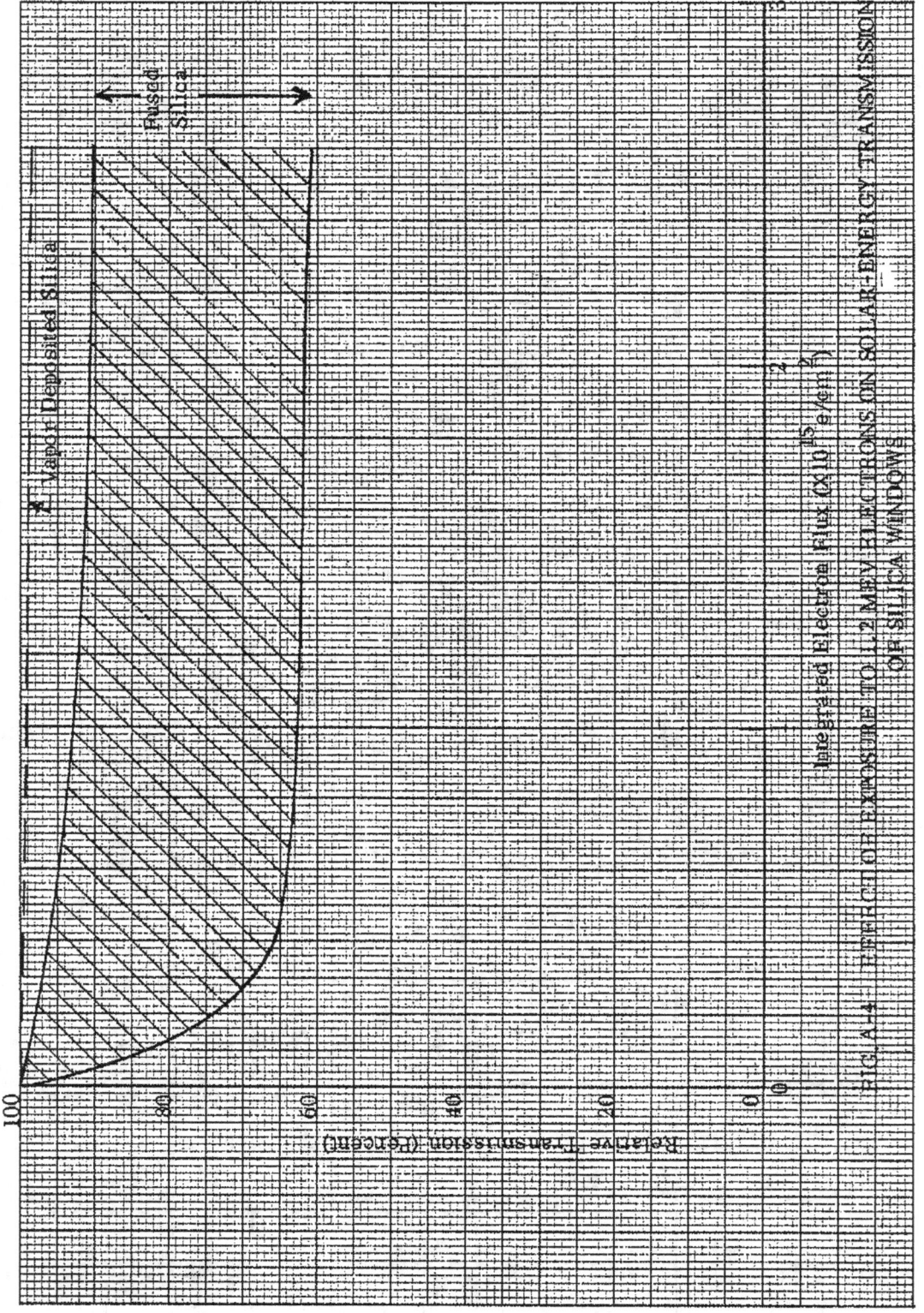

FIG. A-4 EFFECT OF EXPOSURE TO 1.2 MEV ELECTRONS ON SOLAR-ENERGY TRANSMISSION OF SILICA WINDOWS

and 7.6% for a second. The solar cells were 8% efficiency p/n cells and the light source was a lamp with a tungsten filament operating at $2900^\circ K$ with a 1-inch-thick water filter.

Tests of ordinary quartz shielding by Kallander and Weller (Ref. 119) with an integrated flux of 6×10^{15} 1 Mev electrons per cm^2 also showed considerable darkening, which indicates that impure silica is to be avoided.

2. Glass

Typical varieties of glass are very susceptible to radiation damage as shown in Table A-3. The ordinary soda lime plate glass degraded in transmittance by 26.0% with a dose of 2.7×10^{15} 1.2 Mev electrons/cm^2. Some protection is available with the addition of trivalent cerium which tends to neutralize the color centers, thereby reducing the absorption.

3. Sapphire

Artificial sapphire has been used as a cover shield material for solar cells (e.g. on Telstars I and II). The Langley Research Center irradiated a cover shield made of artificial sapphire from the Linde Company, with 1.2 Mev electrons at a total dose of 2.7×10^{15} e/cm^2. With this flux, no decrease in broad band transmittance was observed. This material as shown in Table A-3 was the only material other than vapor-deposited silica that suffered no measurable decrease in transmittance.

4. Adhesives

Adhesives are often utilized to bond the cover shield to the face of the solar cell. The adhesive requires dimensional stability, transparency, high electrical resistivity, long term adhesion and chemical compatibility with the solar cell. The primary concern to date has been with the change in transparency under proton, electron and ultraviolet radiation. The effect of Cobalt 60 gamma radiation on the resistivity of an Epoxy adhesive is reported by Kallander and Weller (Ref. 119). The surface resistivity decreased to 10^9 ohm-cm at a dose of 2×10^8 Roentgens.

F. A. Campbell of the Naval Research Laboratory has measured the change in spectral transmission of adhesives Dow DER-332, Epocast 15-E and Spectrolab ES-10 with electron irradiation. The adhesives were cast 1-2 mils thick in the same method as used to cement cover slides to solar cells. The adhesives ES-10 and 15-E were cast under a 6 mil Corning 7940 fused silica cover and the DER-332 under a 66 mil fused silica cover. The additional shielding did not help the DER-332 and it degraded more than the ES-10 adhesive which was less well protected. The data are tabulated in Table A-1, numb. 10,11 and 12. The ES-10 shows a marked superiority and did not degrade more than 2%. The 15-E showed a decrease in transmission with each successive electron exposure, particularly at the shorter wavelengths. Crazing and delamination were also visible in this specimen which would diffuse the spectrometer light, therefore giving a lower transmissivity reading. However, this diffused light may still be utilized by a solar cell.

E. Conclusion

The materials tested which indicated a superiority by showing minimum darkening with energetic electrons are Corning 7940 of Corning Glass Works, Suprasil II and Optical Grade Silica of Engelhard Industries, Inc., Spectrasil A of Thermal American Fused Quartz Co., GE 104 of General Electric Co., Dynasil (Optical Grade) of Dynasil Corporation, and synthetic sapphire of the Linde Company. The material tested with the least change in transmittance for 4.6 Mev protons was Corning 7940. The adhesive tested with the least change in transmittance under electron irradiation was the Spectrolab ES-10 specimen.

APPENDIX B

SPACE RADIATION EFFECTS ON TRANSISTORS & DIODES

A. Introduction

In addition to solar cells, the semiconductor devices of primary concern with respect to radiation damage are transistors and diodes, which also depend on diffusion of minority carriers for their operation. Primary consideration will be given to those types of transistors and diodes associated with solar cell power systems. In some applications, power regulation components, such as diodes, Zener diodes and transistors, have been located on the solar panels or satellite skin where they are exposed to space radiation with a minimum of shielding. Even when located inside a satellite, it is found that transistors and diodes having gas-filled cases, can be seriously affected.

B. Radiation Damage Mechanisms

The primary radiation effects to transistors and diodes which are of concern in satellites are (1) the surface effects associated with ionization of gas in the device enclosure, and (2) the permanent damage to the minority carrier diffusion length produced by introduction of lattice defects. The effect of injection of electron-hole pairs by ionization in the semiconductor material is of major significance only under extremely high flux conditions, such as the transient neutron and gamma radiation produced by a nuclear explosion, and therefore will not be discussed here.

C. Surface Effects of Radiation

For transistors and diodes which have a gas-filled enclosure, the rather complex effects of ionization of the gas interacting with the surface of the semiconductor can produce changes in performance parameters at radiation doses as low as 10^4 Roentgens. Such a dose may be accumulated in a period of hours to months in the magnetosphere, depending on the amount of shielding. Therefore it is important to select transistors which are least sensitive to this effect.

Reference should be made to the papers by Peck, Blair, Brown and Smits (Ref. 141 or 142) for a detailed discussion of this surface effect which occurs in transistors under a combination of a radiation environment, a bias voltage, and a gas in contact with the surface. The effects observed depend on the junction bias, the envelope voltage, and in many respects on the total radiation dose, rather than dose rate. Upon removal of the bias voltage or the radiation field, a device tends to recover.

It has been found that large variations in test results are observed between units of the same type made by the same manufacturer.

It was found that some devices can be screened for sensitivity to radiation by means of a short-time, high level radiation dose, with correlation to the subsequent behavior under low-level radiation exposure.

The data generally indicate that transistors and diodes in evacuated cases and planar units (with protection over the semiconductor surface) are least affected by ionizing radiation (Ref. 143). By the use of evacuated cases, the dose at which a noticeable increase in the collector reverse current (I_{C0}) occurs can be extended to the range of 10^6 to 10^7 Roentgens as seen in Fig. B-1 (from Ref. 142).

D. <u>Degradation of Diffusion Length in Transistors</u>

Of the changes in semiconductor properties affected by radiation, changes in minority carrier diffusion length (or lifetime) are the predominant cause of permanent damage in transistors and diodes. A large amount of testing has been conducting on germanium and silicon devices in neutron and gamma ray fluxes and damage coefficients have been determined. However, there is no clear correlation which permits the use of data determined in neutron or gamma ray fluxes to predict damage by protons or electrons (Ref. 125).

The transistor parameter which is affected most seriously by a degradation of the diffusion length is the common emitter forward gain (h_{FE}), which is also called β. Bilinski et al (Ref. 125) give the following equations for correlating transistor performance:

$$\frac{1}{h_{FE}} = \frac{1}{h_{FE0}} + \bar{t}\, K_\tau\, \dot{\Phi}$$

where h_{FE0} = initial current gain

h_{FE} = current gain after bombardment

$\bar{t}$ = average transit time of minority carriers in base

$\dot{\Phi}$ = integrated flux

K_τ = lifetime damage constant

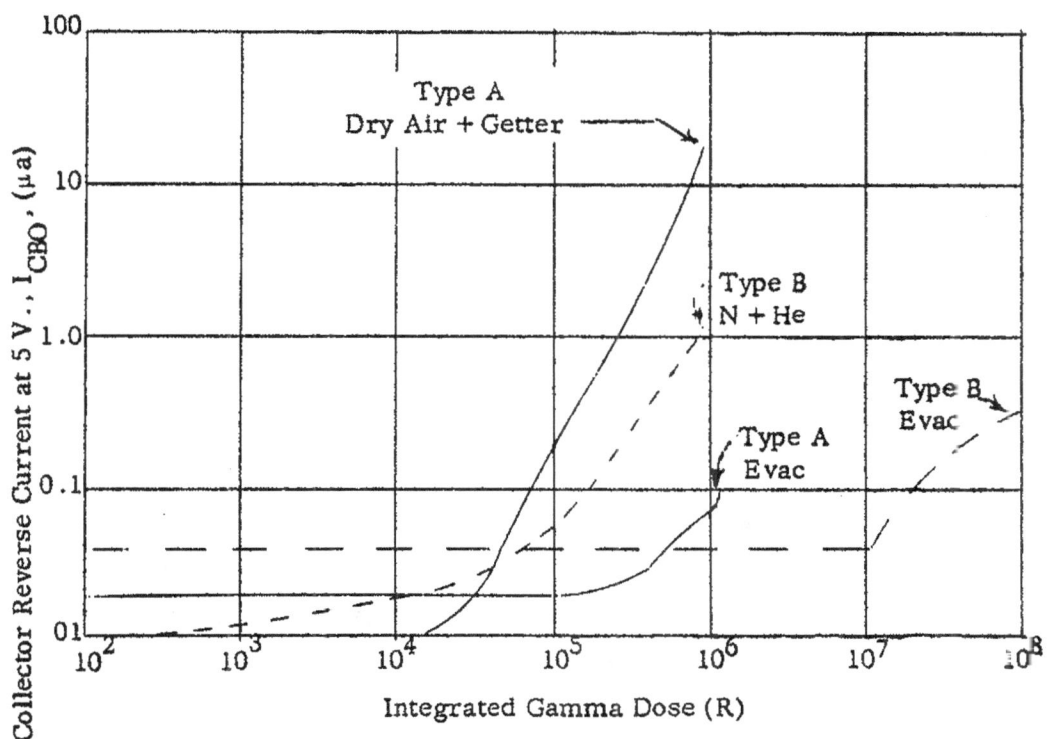

(Radiation Dose Rate 8.5×10^5 rads/hr)

FIG. B-1 EFFECT OF GAMMA RADIATION ON COLLECTOR REVERSE CURRENT OF EVACUATED AND GAS-FILLED SILICON TRANSISTORS

and K_τ is given by

$$\frac{1}{\tau} = \frac{1}{\tau_0} + K_\tau \phi$$

The average transit time of minority carriers in the base region of the transistor is given by:

For a uniform base,
$$\bar{t} = \frac{W^2}{2L} = \frac{1}{2\pi f_{\alpha b}}$$

For a linear graded base
$$\bar{t} = \frac{W^2}{4L} = \frac{1}{\sqrt{2}\pi f_{\alpha b}}$$

where W = base width
L = minority carrier diffusion Length
$f_{\alpha b}$ = alpha cutoff frequency

The above expressions agree well with experiment, except in the case of P-N-P germanium transistors where the effect of depletion layer widening of the collector region is appreciable. (This case has been treated by Easley and Dooley, Ref. 129). Except in this case, the lifetime damage constant for germanium and silicon transistors is given by:

$$K_\tau = (h_{FE0} - h_{FE}) \left(\frac{2\pi f_{\alpha b}}{h_{FE0} h_{FE}}\right)$$

K_τ is related to the diffusion length damage constant K_L by $K_\tau = D K_L$

where D is the diffusion coefficient for minority carriers.

Measurements of germanium and silicon transistor performance after proton irradiation at 22 and 240 Mev have been reported by Hulten, Honaker and Patterson of the NASA Langley Research Center (Ref. 42). Tests with 40 and 440 Mev protons are reported by Honaker and Bryant (Ref. 132).

An attempt to correlate the Langley data on transistors in terms of lifetime damage coefficients for the base materials was made by Bilinski, et al (Ref. 125). However, the data showed considerable scatter, some values for p-type silicon being an order of magnitude smaller than the damage coefficients measured for p-type silicon in solar cells by Bell Telephone Laboratories.

It is apparent that further careful research is required to obtain quantitative data on which to base predictions of the degradation of transistor performance under penetrating proton radiation. However, general rules and guidelines for the selection of transistors can be derived from available test data.

E. Selection of Transistor Type

There is evidence that germanium transistors are more resistant to proton (and neutron) radiation than are silicon transistors which are designed for the same alpha cutoff frequency. Also the sensitivity to radiation damage is much less for transistors with a narrow base width and a high alpha cutoff frequency. There is some evidence that germanium P-N-P types are more resistant than N-P-N types. (This is found to be true under neutron radiation, Ref. 124). Therefore one may conclude that the most radiation resistant transistor should be a P-N-P germanium type with a thin base region and correspondingly high alpha cutoff frequency. In order to minimize surface effects, the transistor should be vacuum encapsulated. If the temperature environment dictates the use of silicon, the cutoff frequency should be high. There does not appear to be a clear distinction between the resistance of P-N-P and N-P-N silicon types, although, on the basis of solar cell experience, one would expect the N-P-N type to be superior under proton and electron irradiation.

Unless care is taken to avoid surface effects, a transistor may fail by increase in collector reverse current at a total radiation exposure much lower than that required to degrade the current gain.

F. Diodes

The behavior of diodes and rectifiers, and Zener reference diodes in neutron and gamma ray fields has been reviewed by Reid (Ref. 64). In general, there was often an initial change in the reverse characteristic of diodes (which was probably attributable to surface effects), followed by a degradation of the forward characteristic.

The life determined by the allowable increase in forward impedance can be greatly increased by using a diode of narrow base width. A factor of 10 reduction in base width yields an increase in radiation tolerance by a factor of about 100.

In comparing alloy and diffused diode structures, Huth (Ref. 133) observes that the diffused structure markedly improves radiation behavior. Reid (Ref. 64) concludes that the optimized configuration of a silicon diode for use in radiation fields would consist of a diffused p - i - n structure with the minimum base thickness consistent with peak inverse - voltage requirements. Based on present knowledge of surface effects, it appears that vacuum encapsulation would be desirable.

G. <u>Shielding Against Space Radiation</u>

In estimating the life of transistors and diodes in a space environment, it must be recognized that surface effects are dependent on the ionization rate produced by protons, electrons and Bremstrahlung (x-rays) in a gas filled device enclosure, whereas the degradation of minority carrier lifetime depends on the flux and energy spectrum of those penetrating electrons and protons which have sufficient energy to produce lattice displacements (> 145 Kev for electrons and > 98 ev for protons in silicon).

The severity of the surface effect problem is critically dependent on the flux and energy spectrum of the electrons in the artificial belt which are predominant over protons in determining the ionization rate (RADS/day) in gas filled devices (Ref. 143). Fortunately, there is evidence that the electron energy spectrum is now much softer (relatively fewer high energy electrons) than a fission spectrum, at least in the upper regions of the artificial belt (Ref. 126). However, it appears desirable to use evacuated devices where possible and to locate any gas-filled devices inside a satellite to take maximum advantage of available shielding, if operation is required in the artificial electron belt. Gas-filled devices should be carefully selected and tested for sensitivity to surface effects.

If transistors or diodes are used outside the satellite shell, they should be evacuated and have protection by at least the same thickness of shielding as the solar cells. For example, at least 0.3 grams/cm^2 of shielding (45 mils of aluminum) is desirable for a satellite orbit like Telstar or Relay which passes through the inner proton and electron belts.

VII REFERENCES

A. References on Radiation Effects on Solar Cells

1. Anderson, R. E. D., Hake, E. A. and Feldman, D., "Telstar Satellite Power System"; Presented at ARS Space Power Systems Conference, Santa Monica, Calif., ARS Paper 2503-62, Sept. 25-28, 1962.

2. Anderson, R. E. D., Meszaros, G. W. and Ciccolella, D. F., "The Satellite Power System (Telstar)"; Bell Laboratories Record, Vol. 41, No. 4, pp. 143-150, April 1963.

3. ASTM Special Technical Publication 330, "Space Radiation Effects on Materials", Published by Amer. Soc. for Testing and Materials, 1916 Race Street, Philadelphia 3, Penn., 1962.

4. Baicker, J. A. and Faughnan, B. W., "Radiation Damage to Silicon Solar Cells"; Summary Report on Contract NAS 5-457, July 31, 1961.

5. Baicker, J. A. and Rappaport, P., "Radiation Damage to Solar Cells"; USAEC Report TID-7652, pp. 118-135, Nov. 1962.

6. Baicker, J. A., Faughnan, B. W. and Wysocki, J. J., "Radiation Damage to Silicon"; Third Semiannual Report on Contract NAS 5-457, RCA Laboratories, Princeton, N. J., April 15, 1963.

7. Baicker, J. A. and Faughnan, B. W., "Radiation-Induced Changes in Silicon Photovoltaic Cells"; J. App. Phys. 33, 3271 (1962).

8. Bell Telephone Laboratories, "NEEP, Nuclear Electronic Effects Program"; Fifth Triannual Technical Note, WADD-TN-61-17, Nov. 15, 1960.

9. Berger, M. J., "Quarterly Status Report - April 1, 1963 through June 30, 1963 on NASA Contract R-80", (Monte Carlo calculation of electron transmission through sapphire shields), National Bureau of Standards, June 30, 1963.

10. Berman, P., "Radiation Damage in Silicon Solar Cells Using 750 Kev and 2 Mev Electrons"; Transitron Electronic Corp. Technical Reprint, presented at Meeting on "Radiation Damage to Semiconductors by High Energy Protons", NASA, Washington, D. C., Oct. 20, 1960.

11. Bilinski, J. R., Brooks, E. H., Cocca, U., Maier, R. J. and Seigworth, D. W., "Proton-Neutron Damage Correlation in Semiconductors"; General Electric Company, Radiation Effects Operation, Syracuse, N. Y., Prepared under Contract NAS 1-1595, June 1962.

12. Billington, D. S. and Crawford, H., Jr., "Radiation Damage in Solids"; Princeton University Press, 1961.

13. Bobone, R. and Merchant, B. W., "Predicting Damage to Silicon Solar Cells Subject to Complex Radiation Spectra"; AIEE Paper No. CP-62-1279, 1962.

14. Brancato, E. L., "Determination of Significant Differences in Solar Cell Experiments"; Presented at IEEE Photovoltaic Specialists Conference, Washington, D. C., April 10-11, 1963.

15. Brown, W. L., "Damage to Semiconductors from Space Radiation", Presented at Amer. Rocket Soc. Space Nuclear Conference, Gatlinburg, Tenn., May 3, 1961.

16. Brown, W. L., "Radiation Damage Effects on Telstar-1", Presented at IEEE Photovoltaic Specialists Meeting, Washington, D. C., April 10-11, 1963.

17. Carter, J. R. and VanAtta, W. K., "Charged Particle Radiation Damage in Semiconductors, VII: Energy Levels of Defect Centers in Electron and Proton Bombarded Silicon", STL Report 8653-6021-KU-000, Feb. 15, 1963.

18. Cherry, W. R., "Solar Cell Radiation Damage with 1 Mev Electrons"; Presented at Radiation Damage Symposium, Goddard Space Flight Center, Greenbelt, Md., Jan. 23, 1963.

19. Cherry, W. R. and Slifer, L. W., "Solar Cell Radiation Damage Studies with 1 Mev Electrons and 4.6 Mev Protons"; NASA Goddard Space Flight Center Report X-636-63-110, May 27, 1963.

20. Cheslow, M. and Kaye, S., "Feasibility Study Toward Development of Radiation Resistant Solar Cell"; Final Report on Contract NAS 7-92, Electro-Optical Systems, Inc., Pasadena, Calif., EOS Report 2080 - Final, Feb. 28, 1963.

21. Davis, Don D., Jr., "Space Environment and Its Effect on Materials"; Proc. of the NASA University Conference on the Science and Technology of Space Exploration, Vol. 2, pp. 439-49, Chicago, Ill., Nov. 1-3, 1962.

22. Denney, J. M. and Pomeroy, D., "Radiation Damage and Transistor Life in Satellites"; Proceedings of the IRE, Vol. 48, No. 5, pp. 950-2, May 1960.

23. Denney, J. M. and Downing, R. G., "Final Report, Charged Particle Radiation Damage in Semiconductors, I: Experimental Proton Irradiation of Solar Cells"; STL Report 8987-0001-RV-000, Sept. 15, 1961.

24. Denney, J. M., Downing, R. B. and Simon, G. W., "Charged Particle Radiation Damage in Semiconductors, III: The Energy Dependence of Proton Damage in Silicon"; STL Report, Oct. 4, 1962.

25. Denney, J. M., Downing, R. G., Lackman, S. R. and Oliver, J. W., "Estimate of Space Radiation Effects on Satellite Solar Cell Power Supplies"; Space Technology Laboratories, Report EM 10-21, MR-13, Oct. 20, 1961.

26. Denney, J. M., et al, "Preliminary Flight Report No. 2, Tetrahedral Research Satellite No. 1"; Space Technology Laboratories, Report 8685-6002-KU-000, Nov. 19, 1962.

27. Denney, J. M., Downing, R. G., Kirkpatrick, M. E., Simon, G. W. and Van Atta, W. K., "Charged Particle Radiation Damage in Semiconductors, IV: High Energy Proton Radiation Damage in Solar Cells"; Contract NAS 5-1851, STL Report 8653-6017-KU-000, Jan. 20, 1963.

28. Denney, J. M., Downing, R. G. and Van Atta, W. K., "Charged Particle Radiation Damage in Semiconductors, V: Effect of 1 Mev Electron Bombardment on Solar Cells"; STL Report 8653-6018-KU-000, Feb. 11, 1963.

29. Downing, R. G. and Denney, J. M., "Lifetime Dependence in Proton Bombarded Silicon on Minority Carrier Density"; STL Report 8653-6016-KU-000, Jan. 18, 1963.

30. Evans, R. D., "The Atomic Nucleus"; McGraw Hill (1955).

31. Evans, R. D., 'Principles for the Calculation of Radiation Dose Rates in Space Vehicles; Arthur D. Little, Inc., Report 63270-05-01, July 1961.

32. Fischell, R. E., "Solar Cell Performance in the Artificial Radiation Belt"; Presented at the ARS Space Power Systems Conference, Santa Monica, Calif., ARS Paper 2578-62, Sept. 1962.

33. Fischell, R. E., "ANNA-1B Solar Cell Damage Experience"; IEEE Photovoltaic Specialists Meeting, Washington, D. C., April 10-11, 1963.

34. Fischell, R. E., "Effect of the Artificial Radiation Belt on Solar Power Systems"; APL Technical Digest, Vol. 2, No. 2, pp. 8-13, Nov.-Dec. 1962.

35. Goetzel, C. G. and Singletary, J. B., "Space Materials Handbook"; Lockheed Missiles and Space Co., Sunnyvale, Calif., Jan. 1962.

36. Hastings, E. C., Jr., "The Explorer XVI Micrometeoroid Satellite, Supplement I, Preliminary Results for the Period Jan. 14, 1963 through Mar. 2, 1963"; NASA TM X-824, April 1963.

37. Hess, W. N., "The Bomb-Produced Radiation Belt"; IEEE Transactions, Vol. NS-10, No. 1, pp. 8-11, Jan. 1963.

38. Hess, W. N., "Neutrons in Space"; NASA TN D-1696, Feb. 1963.

39. Hess, W. N., "The Artificial Radiation Belt Made on July 9, 1962"; NASA TN D-1687, April 1963.

40. Hess, W. N., Personal Communication, June 1963.

41. Hollister, W. L., "Radiation Effects on Electronic Components: An Annotated Bibliography"; Lockheed Missiles and Space Co., SB-63-13, April 1962, U. S. Dept. of Commerce, OTS Report AD 277840, (Price $2.25).

42. Hulten, W. C., Honaker, W. C. and Patterson, J. L., "Irradiation Effects of 22 and 240 Mev Protons on Several Transistors and Solar Cells"; NASA TN D-718, 1961.

43. Hunter, L. P., "Handbook of Semiconductor Electronics"; McGraw-Hill Inc., New York, Second Edition (1962), Section 3.6.

44. Johnson, F. S., "The Solar Constant"; J. Meteorology, Vol. 11, No. 6, p. 431, 1954.

45. Junga, A. and Enslow, G. M., "Radiation Effects on Silicon Solar Cells"; IRE Transactions on Nuclear Science, June 1959.

46. Kallander, J. W. and Weller, J. F., "Radiation Behavior of Electrical Materials and Components for Space Vehicles"; Presented at the Fifth Navy Science Symposium, Annapolis, Md., April 18, 1961.

47. Keller, J. W., Shelton, R. D., Potter, R. A. and Lacy, L., "A Study of the Effect of Geomagnetically Trapped Radiation on Unprotected Solar Cells"; Proc. of the IRE, Vol. 50, No. 11, pp. 2320-27, Nov. 1962.

48. Kleinman, D. A., "Considerations on the Solar Cell"; Bell System Technical Journal, Vol. 40, pp. 85-115, Jan. 1961.

49. Lehr, S. N. and Tronolone, V. J., "The Space Environment and its Effect on Materials and Component Parts"; IRE Transactions on Reliability and Quality Control, Vol. RQC-10, No. 2, pp. 24-37, Aug. 1961.

50. Lehr, S. N., Martire, L. J. and Tronolone, V. J., "Equipment Design Considerations for Space Environment"; Space Technology Laboratories, Redondo Beach, Calif., Report STL TR-9990-6032-RU00, Feb. 1962.

51. Loferski, J. J. and Rappaport, P., Phys. Rev., 111, 432 (1958).

52. Loferski, J. J. and Wysocki, J. J., "Spectral Response of Photovoltaic Cells"; RCA Review, Vol. 22, pp. 38-55, March 1961.

53. Madey, R., "Solar Cell Degradation by Protons in Space"; USAEC Report TID-7652, Book I, pp. 243-259, Nov. 1962.

54. Mandelkorn, G., McAfee, C., Kesperis, J., Schwartz, L. and Pharo, W., "A New Radiation-Resistant, High-Efficiency Solar Cell"; USASRDL Technical Report 2162.

55. Martin, J. H. Teener, J. W. and Ralph, E. L., "Some Effects of Electron Irradiation and Temperature on Solar Cell Performance"; Johns Hopkins University, Applied Physics Lab. Paper CF-3028, May 15, 1963.

56. Oliver, J. W., "Charged Particle Radiation Damage in Semiconductors, II: Minority Carrier Diffusion Analysis in Photovoltaic Devices"; Space Technology Laboratories, Report 8987-0001-RU-001, Feb. 19, 1962.

57. Power Information Center, "Proceedings of the Solar Working Group Conference, 27 and 28 Feb. 1962, Vol. I Radiation Damage to Semiconductor Solar Devices"; Report PIC-SOL 209/2, April 1962.

58. Power Information Center, "Proceedings of the Solar Working Group Conference, 27 and 28 Feb. 1962, Vol. II. Solar Power System Calibration and Testing"; Report PIC-SOL 209/2.1, April 1962.

59. Power Information Center, "Proceedings of the IEEE Photovoltaic Specialists Conference"; Washington, D. C., April 10-11, 1963, (to be published).

60. Prince, M. B. and Wolf, M., "New Developments in Silicon Photovoltaic Devices"; Journal of the British IRE, pp. 583-595, Oct. 1958.

61. Radiation Effects Information Center, "Space Radiation and Its Effects on Materials"; REIC Memorandum 21, Battelle Memorial Institute, Columbus, Ohio, June 30, 1961.

62. Radio Corporation of America, "Radiation Damage to Silicon Solar Cells"; Quarterly Report 1, 2 and 3 and Summary Report on Contract NAS 5-457, RCA Laboratories, July 1, 1961.

63. Radio Corporation of America, Astro-Electronics Division Report AED-1668, "Applied Research Program on High Temperature Radiation-Resistant Solar-Cell Array"; Quarterly Technical Program Report No. 2 on Contract AF33(657)8490, Aug.-Oct. 1962.

64. Reid, F. J., "The Effect of Nuclear Radiation on Semiconductor Devices"; REIC Report No. 10, U. S. Dept. of Commerce, OTS Report PB 171954, (Price $1.00).

65. Reid, F. J., Moody, J. W. and Willardson, R. K., "The Effect of Nuclear Radiation on Semiconductor Materials"; U. S. Dept. of Commerce, OTS Report PB 171955 (Price $1.25).

66. Reynard, D. L., "Measurements of Silicon Solar Cell Degradation in the Van Allen Radiation Environment"; Lockheed Missiles and Space Co., Sunnyvale, Calif., Report LMSC - A073668, Feb. 27, 1962.

67. Rogers, S. C., "Radiation Damage to Satellite Electronic Systems"; IEEE Transactions, Vol. NS-10, No. 1, pp. 97-105, Jan. 1963.

68. Rohrbach, E. J. and Goldstein, H. S., "Radiation vs. Electronic Components"; Machine Design, Vol. 35, pp. 101-4, Jan. 3, 1963.

69. Rosenzweig, W., Smits, F. M. and Brown, W. L., "Energy Dependence of Proton Irradiation Damage in Silicon"; Bull. Am. Phys. Soc. II, Vol. 7, p. 437, Aug. 27, 1962.

70. Rosenzweig, W., "Diffusion Length Measurement by Means of Ionizing Radiation"; Bell System Technical Journal, Vol. 41, pp. 1573-1588, Sept. 1962.

71. Rosenzweig, W., Gummel, H. K. and Smits, F. M., "Solar Cell Degradation Under 1-Mev Electron Bombardment"; Bell System Technical Journal, Vol. 42, pp. 399-414, March 1963.

72. Rosenzweig, W., "Radiation Damage Studies"; IEEE Photovoltaic Specialists Meeting, Washington, D. C., April 10-11, 1963.

73. Ross, Bernd., "Design Criteria for Satellite Power Supplies Using Radiation Resistant Solar Cells"; Presented at ASME Aviation and Space, Hydraulic and Gas Turbine Conference, Los Angeles, Calif., Mar. 3-7, 1963.

74. Shockley, W., "Electrons and Holes in Semiconductors"; Van Nostrand, Inc., New York (1954), p. 309-315.

75. Smits, F. M., Smith, K. D. and Brown, W. L., "Solar Cells for Communication Satellites in the Van Allen Belt"; Journal of the British IRE, pp. 161-169, August 1961.

76. Smits, F. M., "The Degradation of Solar Cells Under Van Allen Radiation"; IEEE Transactions on Nuclear Science, Vol. NS-10, No. 1, pp. 88-96, Jan. 1963.

77. Statler, R. L., "One Mev Electron Damage in Silicon Solar Cells"; IEEE Photovoltaic Specialists Meeting, Washington, D. C., April 10-11, 1963.

78. Statler, R. L., "Solar Cell Radiation Damage with 4.6 Mev Protons"; U. S. Naval Research Lab., Preliminary Report, May 1, 1963.

79. Van Lint, V. A. J. and Wikner, E. B., "Correlation of Radiation Types with Radiation Effects"; IEEE Transactions on Nuclear Science, Vol. NS-10, No. 1, pp. 80-87, Jan. 1963.

80. Van Lint, V. A. J., et al, "Radiation Effects on Silicon Solar Cells"; Final Report on Contract NAS 7-91, General Atomic Division of General Dynamics, Report GA-3872, Feb. 15, 1963.

81. Waddell, R. C., "Radiation Damage Experiments on Relay-1"; IEEE Photovoltaic Specialists Meeting, Washington, D. C., April 10-11, 1963.

82. Waddell, R. C., Goddard Space Flight Center, Personal Communication, June 1963.

83. Weller, J. F., "Proton Damage to Silicon Solar Cells"; NRL Progress, pp. 1-6, Jan. 1963.

84. Weller, J. F. and Statler, R. L., "Low Energy Proton Damage to Solar Cells"; Presented at IEEE Summer General Meeting, Toronto, Canada, June 19, 1963.

85. Wertheim, G. K., "Energy Levels in Electron-Bombarded Silicon"; Phys. Rev. 105:1730, Mar. 15, 1957.

86. Wolf, M., "Limitations and Possibilities for Improvement of Photovoltaic Solar Energy Converters"; Proc. of the IRE, Vol. 48, pp. 1246-1263, 1960.

87. Wysocki, J. J., "The Effect of Series Resistance on Photovoltaic Solar Energy Converters"; RCA Review, pp. 57-70, March 1961.

88. Wysocki, J. J., "Radiation Damage in Solar Cells"; Presented at IEEE Photovoltaic Specialists Conference, Washington, D. C., April 10-11, 1963.

B. References on the Space Environment

89. Allen, L., Jr., Beavers, J. L., II, Whitaker, W. A., Welch, J. A., Jr. and Walton, R. B., "Project Jason Measurement of Trapped Electrons From A Nuclear Device by Sounding Rockets"; Symposium on Scientific Effects of Artificially Introduced Radiations at High Altitudes, Journ. Geophys. Res., Vol. 64, pp. 893-907, August 1959.

90. Brown, W. L., "Trapped Particle Population From Telstar and Explorer XV Observations"; 44th Annual Meeting of American Geophysical Union, Washington, D. C., April 17-20, 1963.

91. Davis, L. R., Williamson, J. M., "Low Energy Trapped Protons"; NASA Publication X-611-62-89, Paper presented at Third International Space Science Symposium and COSPAR Plenary Meeting in Washington, D. C., April 30 - May 9, 1962.

92. Dessler, A. J., "Penetrating Radiation"; Published in "Satellite Environment Handbook", Lockheed Missiles and Space Division, Sunnyvale, Calif., Edited by F. S. Johnson, Report LMSD-895006, Dec. 1960.

93. Frank, L. A., Van Allen, J. A. and Macagno, E., "Charged-Particle Observations in the Earth's Outer Magnetosphere"; Journ. of Geophys. Res., Vol. 68, No. 12, pp. 3543-54, June 15, 1963.

94. Gorchyakov, E. V., "Dispositon of the Inner Radiation Belt and Magnetic Field of the Earth"; AIAA Journal, Vol. 1, No. 2, pp. 520-521.

95. "Goddard Space Flight Center Contributions to the COSPAR Meeting, May 1962"; NASA Publication TND-1669.

96. Hess, W. N., "The Artificial Radiation Belt Made on July 9, 1962"; NASA TND-1687, April 1963.

97. Lehr, S. N., Martire, L. J. and Tronolone, V. J., "Equipment Design Considerations for Space Environment"; STL TR-9990-6032-RU000, Space Technology Laboratories Inc., Redondo Beach, Calif.

98. Lenchek, A. M. and Singer, S. F., "Geomagnetically Trapped Protons from Cosmic Ray Albedo Neutrons"; Journ. of Geophys. Res., Vol. 67, No. 4, pp. 1263-1287, April 1962.

99. McIlwain, C. E., "Coordinates for Mapping the Distribution of Magnetically Trapped Particles"; Journ. of Geophys. Res., Vol. 66, No. 11, pp. 3681-91, Nov. 1961.

100. McIlwain, C. E. and Pizzella, G., "On the Energy Spectrum of Protons Trapped in the Earth's Inner Van Allen Zone"; Journ. of Geophys. Res., Vol. 68, No. 7, pp. 1811-23, April 1, 1963.

101. Naugle, J. E. and Kniffen, D. A., "The Flux and Energy Spectra of the Protons in the Inner Van Allen Belt"; NASA TND-412, 1962.

102. Naugle, J. E. and Kniffen, D. A., "Variations of the Proton Energy Spectrum with Position in the Inner Van Allen Belt"; NASA Goddard Space Flight Center, Preprint X-611-63-35, 1962.

103. O'Brien, B. J., "A Large Diurnal Variation of the Geomagnetically Trapped Radiation"; Journal of Geophys. Res., Vol. 68, No. 4, pp. 989-995, Feb. 15, 1963.

104. O'Brien, B. J., "Review of Studies of Trapped Radiation with Satellite-Borne Apparatus"; Space Science Reviews, Vol. 1, No. 3, pp. 415-484, 1963.

105. O'Brien, B. J., "Radiation Belts"; Scientific American, Vol. 208, No. 5, pp. 84-96, May 1963.

106. Pieper, G. F., "The Artificial Radiation Belt"; APL Technical Digest, Vol. 2, No. 2, pp. 3-7, Nov.-Dec. 1962.

107. Rosen, A., Eberhard, C. A., Farley, T. A. and Vogl, J. L., "A Comprehensive Map of the Space Radiation Environment"; Space Technology Laboratories, Report No. 8644-6002-RU-000.

108. Rosser, W. G. V., "Changes in the Structure of the Outer Radiation Zone Associated with the Magnetic Storm of September 30, 1961"; Journ. of Geophys. Res., Vol. 68, No. 10, pp. 3131-3148, 1963.

109. Walt, M., Crane, G. E. and McDonald, W. M., "Analysis of Atmospheric Scattering Rates for Geomagnetically Trapped Electrons"; 44th Annual Meeting of American Geophysical Union, Washington, D. C., April 17-20, 1963.

110. Van Allen, J. A., McIlwain, C. E. and Ludwig, G., "Satellite Observations of Electrons Artificially Injected into the Geomagnetic Field"; Symposium on Scientific Effects on Artificially Introduced Radiations at High Altitudes, Journ. of Geophys. Res., Vol. 64, pp. 872-91, August 1959.

111. Van Allen, J. A., "Spatial Distribution and Time Decay of the Intensities of Geomagnetically Trapped Electrons from the High Altitude Nuclear Burst of July 1962"; State University of Iowa, Research Report No. 63-11, May 1963.

C. References on Radiation Effects to Cover Materials

112. Campbell, F. J., "Effects of Space Radiation on Solar Cell Cover Materials"; Proceedings of the Photovoltaic Specialists Conference of IEEE, Washington, D. C., April 1963.

113. Campbell, F. J., **Personal Communication, June 1963.**

114. Davis, Don D., "Space Environment and Its Effect on Materials"; Proceedings of the NASA - University Conference on the Science and Technology of Space Exploration, Vol. 2, pp. 439-449, Nov. 1962.

115. Gonshery, Marvin E., "Pulsed Nuclear Radiation Effects on Optical Materials"; Presented at Fall Meeting Optical Soc. of Am., Los Angeles, Calif., Oct. 1961.

116. Hulten, W. C., Honaker, W. C. and Patterson, J. L., "Irradiation Effects of 22 and 240 Mev Protons on Several Transistors and Solar Cells"; NASA Technical Note D-718, April 1961.

117. Hesketh, R. U., "Photon Creation and Destruction of F Centres"; Phil. Mag. 4, 8th Series, pp. 114-125, 1959.

118. Jaffe, L. D. and Rittenhouse, J. B., "Behavior of Materials in Space Environments"; American Rocket Society, 2033-61, Space Flight Report to the Nation - New York, Oct. 9-15, 1961.

119. Kallander, J. W. and Weller, J. F., "Radiation Behavior of Electrical Materials and Components for Space Vehicles"; Fifth Navy Science Symposium, Annapolis, Md., April 1961.

120. Patterson, J. L. and Haynes, G. A., "Effect of High Energy Electron Radiation on Solar Cell Shields"; Private Communication, May 1963.

121. Seitz, F., "Color Centers in Alkali Halide Crystals, II"; Rev. Mod. Phys. 26, pp. 7-94, 1954.

D. References on Radiation Effects to Transistors and Diodes

122. American Society for Testing and Materials, "Space Radiation Effects on Materials"; Committee E-10 on Radioisotopes and Radiation Effects, Technical Publication No. 330, Philadelphia, Pa., p. 43, Oct. 1962.

123. Bertolotti, Mario, "Radiation Damage on Diodes and Transistors"; Alta Frequenza, Vol. 31, No. 2.

124. Bilinski, J. R. and Merrill, R., "Selecting Transistors for Radiation Environments"; Electronics, pp. 38-40, Dec. 25, 1959.

125. Bilinski, et al, "Proton-Neutron Damage Correlation in Semiconductors"; General Electric Co., Radiation Effects Operation, Syracuse, N. Y., Final Report on Contract NAS 1-1595, pp. 32-33, June 1962.

126. Brown, W. L., Personal Communication, July 15, 1963.

127. Clark, J. W., "Effects of Radiation on Semiconductors"; Electronic Industries, Vol. 16, No. 8, p. 80, 1957.

128. Clark, J. W., Wiser, H. L., Petroff, M. D., "Radiation Effects on Silicon Diodes"; Institute of the Radio Engineers - Wescon Convention Record, Vol. 1, No. 43, 1957.

129. Easley, J. W. and Dooley, J. A., "On the Neutron Bombardment Reduction of Transistor Current Gain"; J. App. Phys., Vol. 31, No. 6, pp. 1024-8, June 1960.

130. Freidman, A., Slater, L. M., Alan Kaw, H. F., and Thorpe, L. M., "The Effect of Reactor Radiation and Temperature on Silicon Junction Diodes"; General Electric Cincinnati Report APEX 462, AF 33(600)-38062, Feb. 1959.

131. Gardner, L. B., "Semiconductors and Space Radiation"; Solid State Design Vol. 3, No. 4, pp. 42-46, April 1962.

132. Honaker, W. C. and Bryant, F. R., "Irradiation Effects of 40 and 440 Mev Protons on Transistors"; NASA TND-1490, Jan. 1963.

133. Huth, G. C., "Study Directed Toward Improving the Radiation Tolerance of Silicon Diodes"; Proceedings of the Second AGET Conference on Nuclear Radiation Effects on Semiconductor Devices, Materials and Circuits, Cowan Publ. Corp., Sept. 1959.

134. Keister, G. L. and Stewart, H. V., "The Effect of Radiation on Selected Semiconductor Devices"; Proceedings of the Institute of the Radio Engineers, Vol. 45, pp. 931-937, 1957.

135. Lehr, S. N., Martire, L. J. and Tronolone, V. J., "Equipment Design Considerations for Space Environment"; Space Technology Laboratories Inc., Report STL TR-9990-6032-RU000, Feb. 1962.

136. Levy, G., Touse, R. R. and Castner, S. V., "The Effects of Nuclear Radiation on Some Selected Semiconductor Devices"; Proceedings of the Second AGET Conference on Nuclear Radiation Effects on Semiconductor Devices, Materials and Circuits, Cowan Publ. Corp., Sept. 1959.

137. Loferski, J. J., "Analysis of the Effect of Nuclear Radiation on Transistors"; Journal of Applied Physics, Vol. 29, No. 1, pp. 35-40, Jan. 1958.

138. Loferski, J. J. and Rappaport, P., "Electron Voltaic Study of Electron Bombardment Damage and Its Threshold in Germanium and Silicon"; Phys. Rev., Vol. 98, p. 1861, 1955.

139. Marquardt Corp., "Aircraft Nuclear Propulsion Systems, Project Pluto"; Report 3002, Vol. 3, Contract AF 33(616)-6214, WADC-TN-59-365, Nov. 1959.

140. Miller, W., Bewig, K. and Salzberg, B., "Note on the Reduction of Carrier Lifetime in PN Junction Diodes by Electron Bombardment"; Journal of Applied Physics, Vol. 27, p. 1524, 1956.

141. Peck, D. S., Blair, R. R., Brown, W. L. and Smits, F. M., "Surface Effects of Radiation on Transistors"; The Bell System Technical Journal, Vol. XLll, No. 1, Jan. 1963.

142. Peck, D. S., Blair, R. R., Brown, W. L. and Smits, F. M., "Surface Effects of Radiation on Transistors"; Proceedings of the Symposium on the Protection Against Radiation Hazards in Space, AEC Report TID-7652, Book 1, pp. 136-200, Nov. 1962.

143. Rogers, S. C., "Radiation Damage to Satellite Electronic Systems"; IEEE Transactions on Nuclear Science, Vol. NS-10, No. 1, pp. 97-105, Jan. 1963.

144. Xavier, M. A., "The Performance of Some Zener Reference Elements During Exposure to Nuclear Radiaton"; Third Semiannual Radiation Effects Symposium, Atlanta, Georgia, Oct. 1958.

145. Brown, W. L., Gabbe, J. D. and Rosenzweig, W., "Results of the Telstar Radiation Experiments"; Bell System Technical Journal, July 1963.

146. Mayo, J. S., Mann, H., Witt, F. J., Peck, D. S., Gummel, H. K, and Brown, W. L., "The Command System Malfunction of the Telstar Satellite; Bell System Technical Journal, July 1963.

www.ingramcontent.com/pod-product-compliance
Lightning Source LLC
Chambersburg PA
CBHW081728170526
45167CB00009B/3748